MATHEMATICS
under the Microscope

Notes on Cognitive Aspects
of Mathematical Practice

MATHEMATICS
under the Microscope

Notes on Cognitive Aspects of Mathematical Practice

Alexandre V. Borovik

AMERICAN MATHEMATICAL SOCIETY
Providence, Rhode Island

The image on the cover, "NeuralNet Quilt", is a section of a planar repeating pattern of type pgg and was created by Mike Field, University of Houston. The pattern was generated by a determinstic, but chaotic, torus map, and the colors represent probabilities of visiting points on the pattern under the iteration.

2000 *Mathematics Subject Classification*. Primary 00A30, 00A35, 97C50.

For additional information and updates on this book, visit
www.ams.org/bookpages/mbk-71

Library of Congress Cataloging-in-Publication Data
Borovik, Alexandre.
  Mathematics under the microscope : notes on cognitive aspects of mathematical practice / Alexandre V. Borovik.
    p. cm.
  Includes bibliographical references and index.
  ISBN 978-0-8218-4761-9 (alk. paper)
  1. Mathematics—Psychological aspects. 2. Mathematical ability. I. Title.
BF456.N7B67 2009
510.1′9—dc22

2009029174

À Anna
quand elle était petite fille

Contents

Part I
Simple Things:
How Structures of Human Cognition Reveal Themselves in
Mathematics

Astronomer by Jan Vermeer, 1632–1675. A portrait of Antonij van Leeuwenhoek? Credits: Erich Lessing / Art Resource, NY.

Preface

The portrayal of human thought has rarely been more powerful and convincing than in Vermeer's *Astronomer*. The painting creates the illusion of seeing the movement of thought itself—as an embodied action, as a physical process taking place in real space and time.

I use the *Astronomer* as a visual metaphor for the principal aim of the present book. I attempt to write about mathematical thinking as an objective, real-world process, something which is actually moving and happening in our brains when we do mathematics. Of course, it is a challenging task; inevitably, I have to concentrate on the simplest, atomic activities involved in mathematical practice—hence "the microscope" in the title.

Among other things,

- I look at simple, minute activities, like placing brackets in the sum

$$a + b + c + d + e.$$

- I analyze everyday observations so routine and self-evident that their mathematical nature usually remains unnoticed: for example, when you fold a sheet of paper, the crease for some reason happens to be a perfectly straight line.
- I use palindromes, like MADAM, I'M ADAM, to illustrate how mathematics deals with words composed of symbols—and how it relates the word symmetry of palindromes to the geometric symmetry of solid bodies.
- I even discuss the problem of dividing 10 apples among 5 people!

Why am I earnestly concerned with such ridiculously simple questions? Why do I believe that the answers are important for our understanding of mathematics as a whole?

In this book, I argue that we cannot seriously discuss mathematical thinking without taking into account the limitations of the information-processing capacity of our brains. In our conscious and totally controlled reasoning we can process about 16 bits per second. In activities related to mathematics this miserable bit rate is further reduced to 12 bits per second in the addition of decimal numbers and to 3 bits in counting individual objects. Meanwhile the visual processing module of our brains easily handles 10,000,000 bits per second! (See [211, pp. 138 and 143].) We can handle complex mathematical constructions only because we repeatedly *compress* them until we reduce a whole theory to a few symbols which we can then treat as something *simple*, also because we *encapsulate* potentially infinite mathematical processes, turning them into finite objects, which we then manipulate on a par with other much simpler objects. On the other hand, we are lucky to have some mathematical capacities directly wired into the powerful subconscious modules of our brains responsible for visual and speech processing and *powered* by these enormous machines.

We cannot seriously discuss mathematical thinking without taking into account the limitations of our brains.

As you will see, I pay special attention to *order, symmetry*, and *parsing* (that is, bracketing of a string of symbols) as prominent examples of *atomic* mathematical concepts or processes. I put such "atomic particles" of mathematics at the focus of the study. My position is diametrically opposite to that of Martin Krieger who said in his recent book *Doing Mathematics* [61] that he aimed at

a description of some of the work that mathematicians do, employing *modern and sophisticated* examples.

Unlike Krieger, I write about "simple things". However, I freely use examples from modern mathematical research, and my understanding of "simple" is not confined to the elementary-school classroom. I hope that a professional mathematician will find in the book sufficient non-trivial mathematical material.

The book inevitably asks the question, "How does the mathematical brain work?" I try to reflect on the explosive development of *mathematical cognition*, an emerging branch of neurophysiology which purports to locate structures and processes in the human brain responsible for mathematical thinking [159, 171]. However, I am not a cognitive psychologist; I write about the cognitive mechanisms of mathematical thinking from the position of a practicing mathematician who is trying to take a very close look through the magnifying glass at his own everyday work. I write not so much about discoveries of cognitive science as of their implications for our understanding of mathematical practice. I do not even insist on the ultimate correctness of my interpretations of findings of cognitive psychologists and neurophysiologists. With science developing at its present pace, the current understanding of the internal working of the brain is no more than a preliminary sketch; it is likely to be overwritten in the future by deeper works.

Instead, I attempt something much more speculative and risky. I take, as a working hypothesis, the assumption that mathematics is produced by our brains and therefore bears imprints of some of the intrinsic structural patterns of our minds. If this is true, then a close look at mathematics might reveal some of these imprints— not unlike the microscope revealing the cellular structure of living tissue.

I try to bridge the gap between mathematics and mathematical cognition by pointing to structures and processes of mathematics which are sufficiently non-trivial to be interesting to a mathematician, while being deeply integrated into certain basic structures of our minds and which

Mathematics is the study of mental objects with reproducible properties.

may lie within reach of cognitive science. For example, I pay special attention to *Coxeter Theory*. This theory lies at the very heart of modern mathematics and could be informally described as an algebraic expression of the concept of symmetry; it is named after H. S. M. Coxeter who laid its foundations in his seminal works [336, 337]. Coxeter Theory provides an example of a mathematical theory where we occasionally have a glimpse of the inner working of our minds. I suggest that Coxeter Theory is so natural and intuitive because its underlying cognitive mechanisms are deeply rooted in both the visual and verbal processing modules of our minds. Moreover, Coxeter Theory itself has clearly defined geometric (visual) and algebraic (verbal) components which perfectly match the great visual/verbal divide of mathematical cognition.

However, in paying attention to the "microcosm" of mathematics, I try not to lose the large-scale view of mathematics. One of the principal points of the book is the essential *vertical* unity of mathematics, the natural integration of its simplest objects and concepts into the complex hierarchy of mathematics as a whole.

One of the principal points of the book is the essential vertical *unity of mathematics.*

The *Astronomer* is, again, a useful metaphor. The celestial globe, the focal point of the painting, boldly places it into a cosmological perspective. The Astronomer is reaching out to the Universe—but, according to the widely held attribution of the painting, he is Vermeer's neighbor and friend Antonij van Leeuwenhoek, the inventor of the microscope and the discoverer of the *microcosm*, a beautiful world of tiny creatures which no one had ever seen before. Van Leeuwenhoek also discovered the cellular structure of living organisms, the basis of the unity of life.

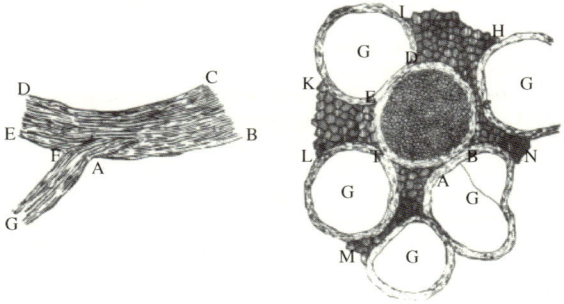

Microstructure of nerve fibers: a drawing by Antonij van Leeuwenhoek, circa 1718. Public domain.

The next principal feature of the book is that I center my discussion of mathematics as a whole—in all its astonishing unity—around the thesis, due to Davis and Hersh [21], that mathematics is

the study of mental objects with reproducible properties.

In this book, the Davis–Hersh thesis works at three levels.

First, it allows us to place mathematics in the wider context of the evolution of human culture. Chapter 11 of the book is a brief diversion into *memetics*, an emerging interdisciplinary area of research concerned with the mechanisms of the evolution of human culture. The term *meme*, an analogue of "gene", was made popular by Richard Dawkins [167] and was introduced into mainstream philosophy and cultural studies by Daniel Dennett [25]. It refers to elementary units of cultural transmission. I discuss the nature and role of "mathematical" memes in detail sufficient, I hope, for making the claim that mathematical memes play a crucial role in many meme complexes of human culture: they increase the precision of reproduction of the complex, thus giving it an evolutionary advantage. Remarkably, the memes may remain invisible, unnoticed for centuries and not recognized as rightly belonging to mathematics. In this book, I argue that this is a characteristic property of "mathematical" memes:

> If a meme has the intrinsic property that it increases the precision of reproduction and error correction of the meme complexes it belongs to and if it does that without resorting to external social or cultural restraints, then it is likely to be an object or construction of mathematics.

So far research efforts in mathematical cognition have been concentrated mostly on brain processes during quantification and counting (I refer the reader to the book *The Number Sense: How the Mind Creates Mathematics* by Stanislas Dehaene [171] for a first-hand account of the study of number sense and numerosity). Important as they are, these activities occupy a very low level in the hierarchy of mathematics. Not surprisingly, the remarkable achievements of cognitive scientists and neurophysiologists are mostly ignored by the mathematical community. This situation may change fairly soon, since conclusions drawn from neurophysiological research could be very attractive to policymakers in mathematics education, especially since neurophysiologists themselves do not shy away from making direct recommendations. I believe that hi-tech "brain scan" cognitive psychology and neurophysiology will more and more influence policies in mathematics education. If mathematicians do not pay attention now, it may very soon be too late; we need a dialogue with the neurophysiological community. The development of neurophysiology and cognitive psychology has reached the point where mathematicians should start some initial discussion of the issues involved. Furthermore, the already impressive body of literature on mathematical cognition might benefit from a critical assessment by mathematicians.

Cognitive psychology and neurophysiology will more and more influence policies in mathematics education. If mathematicians do not pay attention now, it may very soon be too late; we need a dialogue with the neurophysiological community.

Second, the Davis–Hersh thesis puts the underlying cognitive mechanisms of mathematics into the focus of the study.

Finally, the Davis–Hersh thesis is useful for understanding the mechanisms of learning and teaching mathematics: it forces us to analyze the underlying processes of interiorization and reproduction of the mental objects of mathematics.

In my book, I try to respond to the sudden surge of interest in mathematics education which can be seen in the mathematical research community. It appears that it has finally dawned on us that we are a dying breed, that the very reproduction of mathematics as a social institution and a professional community is under threat. I approach the problems of mathematical education from this viewpoint which should not be easily set aside: what kind of mathematics teaching allows for the production of future professional mathematicians? What is it that makes a mathematician? What are the specific traits which need to be encouraged in a student if we want him or her to be capable of a rewarding career in mathematics? I hope that my observations and questions might be interesting to all practitioners and theorists of general mathematical education. But I refrain from any critique of, or recommendations for, school mathematics teaching.

Alexandre Borovik,
aged 11

The *unity* of mathematics means that there are no boundaries between "recreational", "elementary", "undergraduate", and "research" mathematics; in my book, I freely move throughout the whole range. Nevertheless, I try to keep the book as non-technical as possible. I hope that the book will find readers among school teachers as well as students.

In a few instances, the mathematics used appears to be more technical. This usually happens when I have to resort to *metamathematics*, a mathematical description of the structure and role of mathematical theories. But even in such cases, mathematical concepts are no more than a presentation tool for a very informal description of my observations.

Occasionally I could not resist the temptation to include some comments on matters of my own professional interest; however, such comments are indicated in the text by smaller print.

Photographs in this book

> *I come from childhood as from a homeland.*
> Antoine de Saint-Exupéry, *Pilot de guerre*

I tried to place in the margins of the book a photograph of every living mathematician/computer scientist/historian of mathematics/philosopher of mathematics/scholar of mathematics mentioned or quoted in the book. The catch is, I am using *childhood* photographs. In my book, I write a lot about children and early mathematical education, and I wish my book to bear a powerful reminder that we all were children once. I hope that the reader agrees that the photographs make a fascinating gallery—and my warmest thanks go to everyone who contributed his or her photograph.

I tried to place a photograph of a particular person in that section of the book where his/her views had some impact on my writing. The responsibility for my writing is my own, and including a photograph of a person should not be construed as his or her tacit endorsement of my views.

Apologies

> *This book may need more than one preface, and in the end there would still remain room for doubt whether anyone who had never lived through similar experiences could be brought closer to the experience of this book by means of prefaces.*
> Friedrich Nietzsche

I hope that the reader will forgive me that the book reflects my personal outlook on mathematics. To preempt criticism of my sweeping generalizations (and of the even greater sin of using introspection as a source of empirical data), I quote Sholom Aleichem:

Man's life is full of mystery, and everyone tries to compare it to something simple and easier to grasp. I knew a carpenter, and he used to say: "A man is like a carpenter. Look at the carpenter; the carpenter lives, lives and then dies. And so does a man."

And to ward off another sort of criticism, I should state clearly that I understand that, by writing about mathematics instead of doing mathematics, I am breaking a kind of taboo. As G. H. Hardy famously put it in his book *A Mathematician's Apology* [45, p. 61]:

The function of a mathematician is to do something, to prove new theorems, to add to mathematics, and not to talk about what he or other mathematicians have done. Statesmen despise publicists, painters despise art-critics, and physiologists, physicists, mathematicians have similar feelings; there is no scorn more profound, or on the whole justifiable, than that of the men who make for the men who explain. Exposition, criticism, appreciation is work for second-rate minds.

Having broken a formidable taboo of my own tribe, I can only apologize in advance if I have disregarded, inadvertently or through ignorance, any sacred beliefs of other disciplines and professions. To reduce the level of offence, I ask the discerning reader to treat my book not so much as a statement of my beliefs but as a list of questions which have puzzled me throughout my professional career in mathematics and which continue to puzzle me.

Perhaps, my questions are naive. However, I worked on the book for several years and kept the text on the Web, returning to it from time to time to add some extra polish or to correct the errors. So far, the changes in the book were limited to expanding and refining the list of questions, not inserting answers—I cannot find any in the existing literature. This is one of the reasons why I believe that perhaps at least some of my questions deserve a thorough discussion in the mathematical, educational, and cognitive science communities.

My last apology concerns the use of terminology. Some terms and expressions which attained a specialized meaning in certain mathematics-related disciplines are used in this book in their (original) wider and vaguer sense and therefore are more friendly to the readers. To fend off a potential criticism from nit picking specialists, I quote a fable which I heard from one of the great mathematicians of our time, Israel Gelfand:

> A student corrected an old professor in his lecture by pointing out that a formula on the blackboard should contain cotangent instead of tangent. The professor thanked the student, corrected the formula and then added:
> "Young man, I am old and no longer see much difference between tangent and cotangent—and I advise that you do so as well."

Indeed, when mathematicians informally discuss their work, they tend to use a very flexible language—exactly because the principal technical language of their profession is exceptionally precise. I follow this practice in my book; I hope it allows me to be friendly towards all my readers and not only my fellow mathematicians.

Acknowledgements: Inspiration and Help

The gods have imposed upon my writing
the yoke of a foreign tongue
that was not sung at my cradle.
Hermann Weyl

I thank my children, Sergey and Maria, who read a much earlier version of the book and corrected my English (further errors introduced by me are not their responsibility) and who introduced me to the philosophical writings of Terry Pratchett. I am grateful to my wife, Anna, the harshest critic of my book; this book would never have appeared without her. She also provided a number of illustrations.

As the reader may notice, Israel Gelfand is the person who most influenced my outlook on mathematics. I am most grateful to him for generously sharing with me his ideas and incisive observations.

I am indebted to Gregory Cherlin and Reuben Hersh and to my old friend Owl for most stimulating conversations and many comments on the book; some of the topics in the book were included on their advice.

Almost everyday chats with Hovik Khudaverdyan about mathematics and the teaching of mathematics seriously contributed to my desire to proceed with this project.

During our conversation in Paris, the late Paul Moszkowski forcefully put forth the case for the development of the theory of Coxeter groups without reference to geometry and pointed me toward his remarkable paper [388].

Jeff Burdges, Gregory Cherlin, David Corfield, Chandler Davis, Ed Dubinsky, Erich Ellers, Tony Gardiner, Ray Hill, Chris Hobbs, David Pierce, John Stillwell, Robert Thomas, Ijon Tichy, and Neil White carefully read and corrected the whole or parts of the book.

My thanks are due to a number of people for their advice and comments on the specific areas touched upon in the book: to David Corfield—on the philosophy of mathematics, to Susan Blackmore—on memetics, to Vladimir Radzivilovsky—for explaining to me the details of his teaching method, to Satyan Devadoss—on diagrams and drawings used in this book, to Ray Hill—on the history of coding theory, to Péter Pál Pálfy—on universal algebra, to Sergey Utyuzhnikov—on chess, turbulence, and dimensional analysis, to Alexander Jones and Jeremy Gray—on the history of Euclidean geometry, to Victor Goryunov—on multivalued analytic functions, to Thomas Hull—on the history of Origami, to Gordon Royle—on Sudoku, to Alexander Kuzminykh and Igor Pak—on convex geometry, to Dennis Lomas—on visual thinking, to Semen Kutateladze—on philosophy and convex geometry, and, finally, to Paul Ernest and Inna Korchagina for general encouraging comments.

Jody Azzouni, Barbara Sarnecka, and Robert Thomas sent me the texts of their papers [5, 6], [163, 225], [92].

David Petty provided diagrammatic instructions for the Origami Chinese Junk (Figures 11.2 and 11.3). Dougald Dunham allowed me to use his studies of hyperbolic tessellations in M. C. Escher's engravings (Figures 5.4 and 5.5). Ali Nesin made illustrations for Chapter 10. Simon Thomas provided me with diagrams used in Section 12.8.

I am lucky that my university colleagues David Broomhead, Paul Glendinning, Bill Lionheart, and Mark Muldoon are involved in research in mathematical imaging and/or mathematical models of neural activity and perception; their advice has been invaluable.

Paul Glendinning gave me permission to quote large fragments of his papers [183, 185].

My work on genetic algorithms shaped my understanding of the evolution of algorithms; I am grateful to my collaborator Rick Booth who shared with me the burden of the project. Also, the very first seed which grew into this book can be found in our joint paper [108].

Finally, my thanks go to the blogging community—I have picked from the blogosphere some ideas and quite a number of references—and especially to numerous anonymous commentators on my blog.

Anonymous,
age unknown

Acknowledgements: Hospitality

I developed some of the ideas of Section 7.1 in a conversation with Maria do Rosário Pinto; I thank her and Maria Leonor Moreira for their hospitality in Porto.

Parts of the book were written during my visits to University Paris VI in January 2004 and June 2005 on invitation from Michel Las Vergnas, and I use this opportunity to tell Janette and Michel Las Vergnas how enchanted I was by their hospitality.

Section 10.5 of the book is a direct result of a mathematical tour of Cappadocia in January 2006, organized by my Turkish colleagues Ayşe Berkman, David Pierce, and Şükrü Yalçınkaya—my warmest thanks to them for their hospitality in Turkey on that and many other occasions.

Acknowledgements: Institutional

An invitation to the conference *The Coxeter Legacy: Reflections and Projections* at the University of Toronto had considerable influence on my work on this book, and I am most grateful to its organizers.

My work on genetic algorithms was funded by EPSRC (grant GR/R29451).

While working on the book, I used, on several occasions, the facilities of Mathematisches Forschungsinstitut Oberwolfach, The Fields Institute for Research in Mathematical Sciences, and the Isaac Newton Institute for Mathematical Sciences.

Chapter 7 of this book was greatly influenced by the Discussion Meeting *Where will the next generation of UK mathematicians come from?* held in March 2005 in Manchester. The meeting was supported by the Manchester Institute for Mathematical Sciences, by the London Mathematical Society, by the Institute of Mathematics and Applications, and by the UK Mathematics Foundation.

It was during the MODNET Conference on Model Theory in Antalya, November 2–11, 2006, that I placed the first chapter of the book on the Internet. MODNET (Marie Curie Research Training Network in Model Theory and Applications) is funded by the European Commission under contract no. MRTN-CT-2004-512234.

In July 2007 and July 2008 I enjoyed the hospitality of Mathematical Village in Şirince, Turkey, built and run by Ali Nesin.

I started writing this book in *Café de Flore*, Paris—an extreme case of vanity publishing! Since then, I continued my work in many fine establishments, among them *Airbräu, das Brauhaus im Flughafen* in Munich, *Cafe del Turco* in Antalya, *L'authre Bistro* on rue des Ecoles, and *Café des Arts* on place de la Contrescarpe in Paris—I thank them all.

Alexandre Borovik
July 10, 2009
Didsbury

List of Figures

List of Photographs

Part I

Simple Things:
How Structures of Human Cognition
Reveal Themselves in Mathematics

1

A Taste of Things to Come

This is the opening chapter of the book, and I use it to set the tone of my narrative: I start with some simple mathematical observations and briefly discuss what they possibly say about the inner workings of our minds. Surprisingly, this discussion very naturally involves some non-trivial ideas and results from the frontier of mathematical research. But it is better to see it for yourself.

1.1 Simplest possible example

Simplicity, simplicity, simplicity!
I say, let your affairs be as two or three,
and not a hundred or a thousand;
instead of a million count half a dozen,
and keep your accounts on your thumb-nail.

Henry David Thoreau, *Walden*

In my account, I am not afraid to be very personal, almost sentimental, and have decided to start the discussion of the "simple things" of mathematics by turning to my memories from my school years.

I had my most formative mathematical experiences at the tender age of thirteen, when I still lived in my home village on the shores of Lake Baikal in Siberia. I learned elementary calculus from two thin booklets sent to me from a mathematics correspondence school: *The Method of Coordinates* [266] and *Functions and Graphs* [267]. Much later in my life I met one of the authors of the books, the famous mathemati-

Always test a mathematical theory on the simplest possible example—and explore the example to its utmost limits.

3

cian Israel Gelfand, and had a chance to do some mathematics with him.

Once I mentioned to Gelfand that I read his *Functions and Graphs*; in response, he rather sceptically asked me what I had learned from the book. He was delighted to hear my answer: "The general principle of always looking at the simplest possible example". "Yes!" exclaimed Gelfand in his usual manner, "yes, this is my most important discovery in mathematics teaching!" He proceeded by saying how proud he was that, in his famous seminars, he always pressed the speakers to provide simple examples, but, as a rule, he himself was able to suggest a simpler one. [1]

So, let us look at the principle in more detail:

Always test a mathematical theory on the simplest possible example. . .

This is a banality, of course. Everyone knows it; therefore almost no one follows it. So let me continue:

. . . and explore the example to its utmost limits.

This book contains a number of examples pushed to their intrinsic limits. See, in particular, Section 2.6 and the discussion of Figure 2.11 on page 40 for some examples from the theory of Coxeter groups and mirror systems. What could be simpler than that?

But it is even more instructive to look at an example from *Functions and Graphs*.

What is the simplest graph of a function? Of course, that of a linear function,

$$y = ax + b.$$

But what are the simplest non-linear elementary functions? The apparent answer is quadratic polynomials. Well, *Functions and Graphs* suggests something different. The simplest non-linear function is the *magnitude*, or *absolute value*, $y = |x|$.

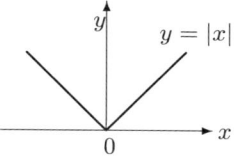

Indeed, it allows

- easy plotting and interpolation;

- standard manipulations with graphs like shifting, stretching, etc.:
$$f(x) + c, \quad f(x + c), \quad cf(x), \quad f(cx).$$

- composition; for example, the composition of $y = |t - 1|$ and $t = |x|$ is
$$f(x) = ||x| - 1|;$$

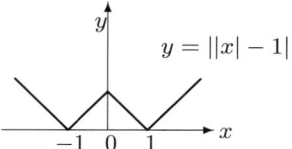

- iterations:
$$f(x) = ||x| - 1|,$$
$$f(f(x)) = |||x| - 1| - 1|,$$
$$f(f(f(x))) = ||||x| - 1| - 1| - 1|,$$
$$\vdots$$

Compare the previous graph of $f(x) = ||x| - 1|$ and the one below of $f(f(x))$:

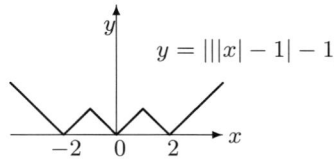

Sketching the 100-th iteration of f becomes an accessible exercise. Can one do the same with polynomials?

And, last but not least, the function $y = |x|$ is not differentiable (and not analytic!), thus providing a simple, natural, and powerful example of a non-analytic function. We shall soon see remarkable implications of this simple observation.

1.2 Switches and flows: some questions for cognitive psychologists

> *I could not fail to disagree with you less.*
> Anonym

One naive notion of function—the one which can eventually be conceptualized as *analytic* function—describes a function as a dependency between two quantities which can be expressed by a *formula*. Historically, this understanding of functional dependence led to the development of the concept of *analytic function*.[2]

However, computations with $y = |x|$ require use of two different formulae for $y < 0$ and $y \geqslant 0$; if you think a few seconds about how you manipulate functions like

$$y = ||||x| - 1| - 1| - 1|,$$

you will see that this is not a symbolic manipulation of the kind we do with analytic expressions, but something quite the opposite, very discrete and consisting almost entirely of flipping, as in bipolar switches LEFT—RIGHT, UP—DOWN.

The difference between the "switch" and "flow" modes of computation is felt and recognized by almost every mathematician.

We humans are apparently quite good at flipping mental switches (when the number of switches is reasonably small; the limitations are possibly of the same nature as in the subitizing/counting threshold; see Section 4.1). Graphic manipulation with compound functions built from $y = |x|$ is so efficient because they appear to engage some small but efficient switchboards in our brains. In contrast, most procedures of formula-based undergraduate calculus obviously follow some smooth, "choiceless" pattern. I would make a wild guess that the choiceless, rearrangement-of-formulae routines of elementary calculus and algebra invoke a type of brain activity which is ruled by rhythm and flow, as in music or reciting of chants. I am more confident in suggesting that, in any case, it is very different from the switch-flipping of discrete mathematics. I base this milder conjecture on the anecdotal evidence that problems which require one to combine the two activities (for example, where the calculations should follow different routes depending on whether the discriminant $\Delta = b^2 - 4ac$ of a quadratic equation $ax^2 + bx + c = 0$ is positive or negative) cause substantial trouble to beginning learners of mathematics, especially if they have not been warned in advance about the hard choices they will face.

I claim that the difference between the "switch" and "flow" modes of computation is felt and recognized by almost every mathematician. Most undergraduate students of mathematics in their second or third year of study can judge—and with a surprising degree of certainty and immediacy in their answers—what kind of mathematics is more suitable for them, discrete or continuous. They just know, even if they have never before given any thought to the issue. Perhaps, we should tell them that there is a difference.[3]

Not being a professional neurophysiologist, I can only conjecture that the two types of mathematical activities should be reflected in two different patterns of brain activity, perhaps even easily noticeable with the help of modern brain scan techniques. Meanwhile, within mathematics itself the two modes of calculation are recognized as being intrinsically different and are analyzed to considerable depth. In the next section I briefly describe the findings of mathematicians.

1.3 Choiceless computation

We choose our joys and sorrows long before we experience them.

Kahlil Gibran

So, we started with the absolute value function $y = |x|$ as an example of "the simplest possible example" and are now moving to a mathematical description of the difference between the "switch" and "flow" modes of computation.

As I will frequently do in this book, I use a concept from computer science as a pointer to possible structures of human cognition responsible for particular ways of manipulating mathematical objects. In this case, a possible indicator is the concept of *choiceless polynomial time computation* [313].

Some terminology ought to be explained.

1.3.1 Polynomial time complexity

An algorithm is said to have *polynomial time* complexity (of degree d) if, when working with inputs of size l, it requires $O(l^d)$ elementary operations (see the endnote [4] for an explanation of $O(\ )$-notation). Let us look, for example, at the addition of two integers. The input size here is the number of digits required to write the integers down; if both summands are smaller than n, then each needs at most

$$l = [\log_{10}(n)] + 1$$

digits (here, $[\log_{10}(n)]$ denotes $\log_{10}(n)$ rounded down to the nearest integer). To add the integers, we need, in each position, to add two digits, compute the carry, if necessary, and add it to the next position. All that requires at most 3 operations, making it

$$3([\log_{10}(n)] + 1)$$

in total. It is convenient to say succinctly that the time complexity of addition is

$$O(\log_{10}(n)).$$

Notice further that a change of the basis of the logarithm amounts to multiplication of its value by a constant; hence we can use any basis (actually, computer scientists prefer base 2, since everything is done in binary numbers) and just write $O(\log n)$.

Similarly, the multiplication of two integers has time complexity at most $O(\log^2 n)$, or *quadratic time* complexity, while the multiplication of two $n \times n$ matrices with integer entries, each smaller than m, requires at most $O(n^3 \log^2 m)$ operations. I do not consider here various interesting and useful ways to *speed up* the standard procedures; see the classical book by Knuth [372] for a very comprehensive survey of the huge body of knowledge on fast practical algorithms.

Why is the class $\mathcal{P}$ of polynomial time algorithms so important? From a practical point of view, many known polynomial time algorithms cannot be used on modern computers now or in the near future, since the constant in $O(\ )$, or the power of the polynomial involved, may be too big for the computation to be completed in any feasible amount of time.

As frequently happens in mathematics, $\mathcal{P}$ is being studied because it is robust. The actual degree d in the bound $O(n^d)$ for complexity depends on data structures and other computer science tricks used; but if an algorithm can be implemented as polynomial time one way, it remains polynomial time in all reasonable implementations, and this allows us to work at the theoretical level and ignore the details.

Also, there is a school of thought holding that the existence of a polynomial time algorithm should normally suggest the existence of a good and practically feasible polynomial time algorithm. This is how Neal Koblitz (one of the founders of algebraic cryptography) put it in [373, p. 37]:

> *The experience has been that if a problem of practical interest is in $\mathcal{P}$, then there is an algorithm for it whose running time is bounded by a* small *power of the input length.*

Koblitz's statement is a metamathematical thesis describing a big class of especially nice algorithms.[5] In the next section, I will try to describe even nicer algorithms—but, unfortunately, these are less versatile and less applicable than polynomial time algorithms.

1.3.2 Choiceless algorithms

For our discussion of the "switch" and "flow" modes of computation, we have to move to some firm mathematical ground. Luckily, the necessary formal concept of "choiceless" computation is already well known in computer science.

> *Choiceless computing imposes the black-and-white vision of mathematics. But maybe the same is true in life—if one tries to avoid choices, one sees the world in just two colors.*

In naive terms, a choiceless algorithm is a routine which works in a single uninterrupted flow, never encountering a choice between two (or more) distinct ways to continue the computation.[6]

So, what does mathematics say about choiceless algorithms and their limitations?

A benchmark problem in the theory of choiceless computation is that of evaluation of the determinant of an $n \times n$ matrix. The standard algebraic formula for the determinant

$$\det A = \sum_{\sigma}(-1)^{\operatorname{sign}\sigma} a_{1,\sigma(1)} \cdots a_{n,\sigma(n)}$$

gives a choiceless algorithm working in $O(n!)$—since the sum is taken over all $n!$ permutations of indices $1, \ldots, n$. Therefore, it does not work in polynomial time. In contrast, the standard Gaussian elimination procedure works in cubic time but requires making choices; the first choice is needed at the very first step, when we have to decide whether to swap rows of the matrix (a_{ij}) according as $a_{11} = 0$ or not.

A remarkable result of Blass, Gurevich and Shelah, [314, Theorem 29] asserts that the determinant of a matrix over the field of two elements *cannot* be found in choiceless polynomial time. In ordinary language, their result says that if the determinant is to be calculated efficiently, then choices are essential and unavoidable.

Probably the most advanced result in the theory of choiceless algorithms is Shelah's zero-one law [315, 402] which says that if a property of a class of objects is recognizable in choiceless polynomial time (that is, given an object, we can decide choicelessly and in polynomial time whether the object has the property in question or not), then the property holds with probability 0 or 1. Paradoxically, choiceless computing imposes the black-and-white vision of mathematics. But maybe the same is true in life—if one tries to avoid choices, one sees the world in just two colors, and one color is always preferred.

When using computer science as a source of metaphors for describing the workings of our brain, it is worth remembering that, despite the immense complexity and power of the brain, the mental processes of mathematics appear to be surprisingly resource-limited. Therefore I have a feeling that branches of logic developed for the needs of complexity theory might provide better metaphors than the general theory of computation.

1.4 Analytic functions and the inevitability of choice

AEROFLOT *flight attendant: "Would you like a dinner?"*
Passenger: "And what's the choice?"
Flight attendant: "Yes—or no."

We have mentioned at the beginning of our discussion that $|x|$ is a non-analytic function. It can be written by a single algebraic formula

$$|x| = \sqrt{x^2},$$

with the only glitch being that of the two values of the square root $\pm\sqrt{x^2}$ we have *to choose* the positive one, namely, $\sqrt{x^2}$.

One may argue that in the case of the absolute value function the choice is artificial and is forced on us by the function's awkward definition. But let us turn to solutions of algebraic equations, which give more natural examples of the inevitability of choice.

The classical formula

$$x_{1,2} = \frac{-b \pm \sqrt{b^2 - 4ac}}{2a}$$

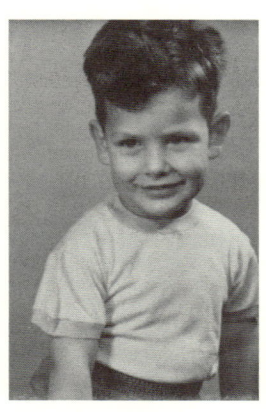

Chris Hobbs,
aged 6

for the roots of the quadratic equation is the limit of what we can do with analytic functions without choosing branches of multivalued analytic functions—but even here, beware of complications and read an interesting comment from Chris Hobbs.[7] Recall that the inverse of the square function $x = y^2$ is a two-valued function $y = \pm\sqrt{x}$ whose graph has two branches, positive $y = \sqrt{x}$ and negative $y = -\sqrt{x}$. Similarly, the cube root function $y = \sqrt[3]{x}$ has three distinct branches, but they become visible only in the complex domain, since only one cube root of a real number is real; the other two are obtained from it by multiplying it by complex factors

$$-\frac{1}{2} \pm \frac{\sqrt{3}}{2}i.$$

The classical formula—which can be traced back to Gerolamo Cardano (1501–1576) and Niccolò Tartaglia (1499–1557)—for the roots of the cubic equation

$$x^3 + ax^2 + bx + c = 0$$

gives its three roots as

$$-\frac{a}{3} + \sqrt[3]{\frac{-2a^3 + 9ab - 27c + \sqrt{(2a^3 - 9ab + 27c)^2 + 4(a^2 + 3b)^3}}{54}}$$
$$+ \sqrt[3]{\frac{-2a^3 + 9ab - 27c - \sqrt{(2a^3 - 9ab + 27c)^2 + 4(a^2 + 3b)^3}}{54}},$$

$$-\frac{a}{3} + \frac{-1 - i\sqrt{3}}{2}\sqrt[3]{\frac{-2a^3 + 9ab - 27c + \sqrt{(2a^3 - 9ab + 27c)^2 + 4(a^2 + 3b)^3}}{54}}$$
$$+ \frac{-1 + i\sqrt{3}}{2}\sqrt[3]{\frac{-2a^3 + 9ab - 27c - \sqrt{(2a^3 - 9ab + 27c)^2 + 4(a^2 + 3b)^3}}{54}},$$

$$-\frac{a}{3} + \frac{-1 + i\sqrt{3}}{2}\sqrt[3]{\frac{-2a^3 + 9ab - 27c + \sqrt{(2a^3 - 9ab + 27c)^2 + 4(a^2 + 3b)^3}}{54}}$$
$$+ \frac{-1 - i\sqrt{3}}{2}\sqrt[3]{\frac{-2a^3 + 9ab - 27c - \sqrt{(2a^3 - 9ab + 27c)^2 + 4(a^2 + 3b)^3}}{54}}.$$

Please notice the carefully choreographed choice of the branches of the square root $\sqrt{\ }$ and the cube root function $\sqrt[3]{\ }$, the rhythmic dance of pluses and minuses. Without that choice, Cardano's formula produces too many values, only three of which are true roots.

Indeed, if we work with multivalued functions without making any distinction between their branches, we have to accept that the superposition of an m-valued function and an n-valued function has mn values. We cannot collect like terms: an innocent looking expression like

$$\sqrt{x} + \sqrt{9x}$$

defines, if we interpret "$\sqrt{x}$" as two-valued, a function with *four* branches

$$\pm\sqrt{x} \pm \sqrt{9x} = \{\, -4\sqrt{x}, -2\sqrt{x}, 2\sqrt{x}, 4\sqrt{x} \,\}.$$

It is a rigorous mathematical fact [307] that solutions of equations of degree higher than two cannot be analytically expressed by choiceless multivalued formulae (even if we allow for more sophisticated analytic functions than radicals); see a discussion of the topological nature of this fact by Vladimir Arnold [3, p. 38].

This last observation is especially interesting in the historic context. At the early period of development of symbolic algebra, mathematicians were tempted to introduce functions more general than roots. The following extract from Pierpaolo Muscharello's *Algorismus* from 1478 is taken from Jens Høyrup [54]:

Pronic root is as you say, 9 times 9 makes 81. And now take the root of 9, which is 3, and this 3 is added above 81, so that the pronic root of 84 is said to be 3.

In effect, Muscharello wanted to introduce the inverse of the function

$$z \mapsto z^4 + z.$$

Arnold's theorem explains why such tricks could not lead to an easy solution of cubic and quadric equations and why it had been abandoned.

1.5 You name it—we have it

This section is more technical and can be skipped.

As I have already said on several occasions, this book is about simple atomic objects and processes of mathematics. However, mathematics is huge and immensely rich; even the simplest observations about its simplest objects may already have been developed into sophisticated and highly specialized theories. Mathematics' astonishing cornucopian richness and its bizarre diversity are not frequently mentioned in works on philosophy and methodology of mathematics—but this point has to be emphasized, since its makes the question about *unity* of mathematics much more interesting.

In this section, I will briefly describe a "mini-mathematics", a mathematical theory concerned with a close relative of the absolute value function, the *maximum* function of two variables

$$z = \max(x, y).$$

Of course, the absolute value function $|x|$ can be expressed as

$$|x| = \max(x, -x).$$

Similarly, the maximum $\max(x, y)$ can be expressed in terms of the absolute value $|x|$ and arithmetic operations—I leave it to the

| *Oh yes, do it.* |

reader as an exercise. [?]

The theory is known by the name of *tropical mathematics*. The strange name has no deep meaning: the adjective "tropical" was coined by French mathematicians in honor of their Brazilian colleague Imre Simon, one of the pioneers of the new discipline. Tropical mathematics works with the usual real numbers but uses only two operations: addition, $x + y$, and taking the maximum, $\max(x, y)$—therefore it is one of the extreme cases of "switch-flipping", choice-based mathematics. Notice that addition is distributive with respect to taking maximum:

$$a + \max(b, c) = \max(a + b, a + c).$$

This crucial observation is emphasized by renaming the two basic operations into new "multiplication" and "addition":

$$a \odot b = a + b, \qquad a \oplus b = \max(a, b).$$

The previous identity takes the more familiar shape of the distributive law:

$$a \odot (b \oplus c) = a \odot b \oplus a \odot c.$$

Of course, operations $\odot$ and $\oplus$ are commutative and associative. After recycling the traditional shorthand

$$x^n = x \odot \cdots \odot x \quad (n \text{ times}),$$

we can introduce polynomials as well as matrix multiplication, determinants, etc. For example, the tropical determinant of the matrix $A = (a_{ij})$ is defined by evaluating the expansion formula (see page 9) tropically and ignoring the signs of permutations σ involved in the classical formula for determinant [396, 408]:[8]

$$\det{}_{\mathrm{tr}} A = \bigoplus_{\sigma \in \mathrm{Sym}_n} (a_{1,\sigma(1)} \odot \cdots \odot a_{n,\sigma(n)})$$

$$= \max_{\sigma \in \mathrm{Sym}_n} (a_{1,\sigma(1)} + \cdots + a_{n,\sigma(n)}).$$

Here the "sum" is taken over the set Sym_n of all permutations of indices $1, \ldots, n$; for example,

$$\det{}_{\mathrm{tr}} \begin{vmatrix} a & b \\ c & d \end{vmatrix} = \max(a + d, b + c).$$

Therefore the "sum" involves $n!$ "monomials"—exactly as in the case of ordinary determinants. Let us return for a second to our discussion of the complexity of the evaluation of determinants in two models of computation: choice-based and choiceless. It is not obvious at all that a tropical determinant can be evaluated using $O(n^3)$ elementary operations—but it is true. The evaluation of the tropical determinant is the classical *assignment problem* in discrete optimization and the bounds for its complexity are well known [400, Corollary 17.4b]. Indeed, finding the value of the tropical determinant amounts to finding a permutation $i \mapsto \sigma(i)$ which maximizes the sum

$$a_{1,\sigma(1)} + \cdots + a_{n,\sigma(n)}.$$

This is an old problem of applied discrete optimization: a company has bought n machines $M_1, \ldots, M_n$ which have to be assigned to n factories $F_1, \ldots, F_n$; the expected profit from assigning machine M_i to factory F_j is a_{ij}; find the assignment

$$M_i \mapsto F_{\sigma(i)}$$

which maximizes the expected profit

$$a_{1,\sigma(1)} + \cdots + a_{n,\sigma(n)}.$$

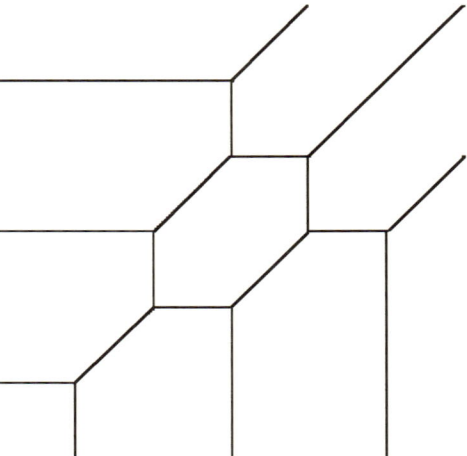

Fig. 1.1. A cubic curve in the tropical projective plane. Mikhalkin [385], reproduced with permission.

We also have a full-blown tropical algebraic geometry, where curves in the plane are made from pieces of straight lines (Figure 1.1)—quite like the graph of the absolute value function $y = |x|$, the starting point of our discussion.

Tropical mathematics is amusing, but is it relevant? Yes, and very much so. Moreover, it currently is experiencing an explosive growth. There are intrinsic mathematical reasons for tropical mathematics to exist, but its present flourishing is largely motivated by applications.

One application is mathematical genomics: tropical geometry captures the essential properties of "distance" between species in the phylogenic tree.

Another is theoretical physics: tropical mathematics can be treated as a result of the so-called Maslov dequantization of traditional mathematics over numerical fields as the Planck constant $\hbar$ tends to zero taking imaginary values [380].

A third application is computer science and the theory of time-dependent systems, like queuing networks (where tropical mathematics is known under the name of a $(\max, +)$-*algebra*. The rationale behind this class of applications is an observation so simple and banal that it has a certain ironic flavor. Indeed, we do not normally multiply time by time; instead, we either add two intervals of time (which corresponds to consecutive execution of two processes) or compare the lengths of two intervals—to decide which process ends earlier. Therefore tropical mathematics is mathematics of *time*—which also explains its applications to genomics: phylogenic trees grow in time, and the geometry of phylogenic trees reflects the geometry of time.

1.6 Why are certain repetitive activities more pleasurable than others?

> *Ivan crossed it all out and decided*
> *to begin right off with something very strong,*
> *in order to attract the reader's attention at once,*
> *so he wrote that a cat had got on a tram-car, and*
> *then went back to the episode with the severed head.*
>
> Michael Bulgakov, *The Master and Margarita*

Let us turn our attention to the emotional side of mathematics, more specifically, to the personal psychological experience of people working with mathematical algorithms and routines.

I wish to formulate here some of my observations and conjectures which may appear to be bizarre and out of tune from the usual discourse on mathematics. However, I tested some of them in a warm-up talk that I gave at the forum discussion *Where do mathematicians come from?* [457], part of a very peculiar conference, that of the World Federation of National Mathematics Competitions (WFNCM). It was held in July 2006 in Cambridge, England. On my way from Manchester to Cambridge, four hours by train, I had seen three people solving Sudoku puzzles. In one case, a lady of middle age shared a table with me and I had a chance to watch, in all detail and with a growing fascination, how she was solving an elementary level Sudoku puzzle. Her actions followed a certain rhythm: first she inspected the puzzle row by row and column by column until she located a critical cell (whose value had been already uniquely determined by the already known values in other cells) and then, with obvious agitation, checked that was indeed the case, happily wrote the digit in, smiled with a childish satisfaction, relaxed for a few seconds, and, after a short pause, started the search again.

The next day, in my talk at the conference, I pointed out that, from a mathematical point of view, solving an elementary level Sudoku puzzle is nothing more than solving a triangular system of Boolean equations by back substitution, something very similar to what we do after a Gauss–Jordan elimination in a system of simultaneous linear equations. But has anyone ever seen people on a train solving systems of linear equations from a newspaper?[9]

Why is Sudoku popular, when systems of linear equations are not? (Actually, I was slightly wrong: at the time of my talk, I was unaware of Kakuro, which combines linear and Boolean equations. But one still has to see whether Kakuro beats Sudoku in popularity.)

We will not understand the psychological and neurophysiological roots of an important aspect of mathematical practice until we figure out why bubble wrap popping is such an addictive and pleasant activity.

Still, why is Sudoku popular? I believe the answer is in a rhythm of repeated cycles of operations each of which engages our brains just up to a right and most pleasurable level of intensity. As a student, I experienced the soothing, relaxing effect of carrying out a recursive algorithm, like long division, or Euclid's algorithm. Later, in my research work, I felt a similar emotional impact from inductive arguments in finite group theory: you start with a minimal counterexample to the theorem and then simplify it step by step, like removing layers from an onion, until you pinpoint the core contradiction and destroy the counterexample. My teacher Victor Danilovich Mazurov expressed the principle of a "minimal counterexample" using a line from a Russian fairy tale:

> The oldest brother hid behind the back of the younger one, the younger one hid behind the youngest one, and the youngest brother fell on his knees, raised his hands and pleaded for mercy.

In mathematical education, especially at its earlier stages, one of the teacher's tasks is to give his/her students the opportunity to feel this soothing, comforting effect of a rhythmic repetitive activity. And here I come to the crucial point:

> why do people love to pop bubble wrap?

I would not write this now if the audience of my talk at the WFNMC conference had not immediately agreed with, and approved of, my comparison of the execution of certain types of recursive algorithms with bubble wrap popping. I should perhaps explain that the audience included some of the best experts on mathematical education in the world, especially on advanced and non-standard aspects of mathematics teaching. They definitely knew everything about the so-called "recreational mathematics", puzzles, brainteasers, and conundrums of every possible kind. Their support allows me to be quite confident in my comparison of Sudoku with bubble wrap popping. In any case, the lady on the train was doing her Sudoku in an immediately recognizable bubble wrap popping rhythm.

So, with the authority of the conference on my side, I dare to formulate my thesis:

> We shall not understand the psychological and neurophysiological roots of an important aspect of mathematical practice until we figure out why bubble wrap popping is such

an addictive and pleasant activity. Why does it comfort and help to relax? Why is it soothing?

Actually, some years ago I formulated a rather embarrassing conjecture that the attraction to bubble wrap popping is genetically determined. Bubble wrap triggers in humans archaic instincts linked to an ape-like behavior: grooming (and even more importantly, mutual grooming) and destruction of lice. In apes and monkeys, mutual grooming is an important part of social bonding, which explains its soothing, comforting, relaxing effect.

In my search on the Web for a confirmation of my conjecture I have not managed to get further than numerous websites devoted to virtual bubble wrap popping. A search on the words "bubble wrap" is not the best way to find anything meaningful on the Web: almost everything sold on the Internet is mailed in bubble wrap packaging. As a result, GOOGLE produces 9,090,000 hits for "bubble wrap". I offered the problem to my colleague Gregory Cherlin, who was more Internet savvy and carried out a successful search. Here are his principal findings:

- Gene HOXB8 controls normal grooming behavior. Disruption in mice leads to obsessive grooming behavior. Here is a summary of information from the National Institutes of Health website [448]:

  This gene belongs to the homeobox family of genes. The homeobox genes encode a highly conserved family of transcription factors that play an important role in morphogenesis in all multicellular organisms. Mammals possess four similar homeobox gene clusters, HOXA, HOXB, HOXC and HOXD, which are located on different chromosomes and consist of 9 to 11 genes arranged in tandem. This gene is one of several homeobox HOXB genes located in a cluster on chromosome 17. HOXB8 knockout mice exhibit an excessive pathologic grooming behavior, leading to hair removal and self-inflicted wounds at overgroomed sites. This behavior is similar to the behavior of humans suffering from the obsessive-compulsive spectrum disorder trichotillomania.

- There is quite a range of grooming-related disorders in humans [454].
- Primates indeed do love to pop bubble wrap [445, p. 8].

Meanwhile, my own search for bubble wrap popping on Google Scholar led me to the book under the telling title *Teens Together Grief Support Group Curriculum* [429]. I have not seen the whole book, but, apparently, page 57 contains sufficiently revealing words:

Bubble Wrap: Give the teens a square of bubble wrap to pop for one of their breaks. They really get into the sound and action of popping the bubbles.

Should we be surprised if it were confirmed indeed that the most comfortable pace of execution of a recursive algorithm is set by a gene responsible for grooming behavior?

As I suspected, the soothing and comforting effect of bubble wrap popping is indeed well known to practicing psychotherapists.

I would suggest that HOXB8 is indeed the Bubble Wrap Gene and is responsible for Sudoku being attractive to humans. I would rather hear more on that from geneticists and neurophysiologists.

At last I am in a position to formulate the moral of this story. I believe that a real understanding of one of the key issues of mathematical practice (and especially of mathematics teaching) cannot be achieved without answering a question:

- why are some objects, concepts, and processes of mathematics more intuitive, "natural", or just more convenient and acceptable than others?

We cannot answer it withou taking a hard and close look at the very deep and sometimes archaic levels of the human mind and the human neural system. Indeed, Stanislas Dehaene said in his book *The Number Sense* [171]:

> *We have to do mathematics using the brain which evolved 30,000 years ago for survival in the African savanna.*

In particular, should we be surprised if it were confirmed indeed that the most comfortable pace of execution of a recursive algorithm is set by a gene responsible for grooming behavior?

1.7 What lies ahead?

We have seen how the deceptively simple function $y = |x|$ launched us on a roller coaster ride through several branches of mathematics. More adventures still lie ahead. They all will follow a similar plot:

- Usually I start by describing a very simple—sometimes ridiculously simple—mathematical problem, object, or procedure.
- Then I discuss possible neurophysiological mechanisms which might underpin the way we think and work with this object. Sometimes my conjectures are purely speculative, sometimes (for example, in the next chapter) they are based on established neurophysiological research.

- I usually include a brief description of mathematical results which deal with *mathematical* analogues of the conjectural neurophysiological mechanisms.
- In the later parts of the book, I will more and more frequently venture into the discussion of possible implications of our findings for our understanding of mathematics and its philosophy.
- My conjectures are frequently outrageous and sketchy. I have no qualms about that. The aim of the book is to ask questions, not give answers.

I very much value a "global" outlook at mathematical practice (in recent books best represented by David Corfield's *Towards a Philosophy of Real Mathematics* [16]), but, in this book, I prefer to concentrate on the "microscopic" level of study.

Quite often the mathematics discussed or mentioned in this book is very deep and belongs to mainstream mathematical research, either recent or, if we talk about the past, of some historic significance. I believe that this is not a coincidence. Mathematics is produced by our brains, which imprint onto it some of the structural patterns of the intrinsic mechanisms of our mind. Even if these imprints are not immediately obvious to individual mathematicians, they are very noticeable when mathematics is viewed on a larger scale—not unlike hidden structures of landscape which emerge in photographs made from a plane or a satellite.

David Corfield,
aged 10

Notes

[1] SIMPLEST POSSIBLE EXAMPLES. Of course, simplest, in the relative sense, examples can be found at every level of mathematics. Here is one example, due to Gelfand: the simplest non-commutative Lie group is the group of isometries of the real line $\mathbb{R}$; it is the extension of the additive group $\mathbb{R}^+$ by the multiplicative group $\{-1, +1\}$. Its representation theory is a well-known chapter of elementary mathematics, namely, trigonometry; however, the connection between representation theory and trigonometry is not frequently discussed. But this is not the simplest possible example of a simplest possible example, and its discussion will lead us beyond the scope of this book.

However, it would be useful to record one consequence of the relation between representation theory and trigonometry: the formula for matrix multiplication is more fundamental than almost any trigonometric formula. We shall return to that later; see page 189.

[2] ANALYTIC FUNCTIONS. A function $f(x)$ is *analytic* at $x = a_0$ if we can write $f(a_0 + z)$ as a power series

$$f(a_0 + z) = a_0 + a_1 z + a_2 z^2 + a_3 z^3 + \cdots$$

converging for all sufficiently small z. For example, the square root function $y = \sqrt{x}$ is analytic at the point $x = 1$ since by Newton's binomial formula we have

$$\sqrt{1+z} = 1 + \frac{z}{2} - \frac{z^2}{4 \cdot 2!} + \frac{3 \cdot z^3}{8 \cdot 3!} - \frac{3 \cdot 5 \cdot z^4}{16 \cdot 4!} + \frac{3 \cdot 5 \cdot 7 \cdot z^5}{32 \cdot 5!} + \cdots$$

for all z such that $|z| < 1$. But a power series expansion for $y = \sqrt{x}$ at $x = 0$ does not exist: the function $y = \sqrt{x}$ is not analytic at $x = 0$.

[3]DISCRETE VS. CONTINUOUS: the two other great divides in mathematics are between "finite" and "infinite" and between "geometric" and "formula-based"; we shall discuss them later in the book. It is quite common to associate the "geometric" or visual, and "formula-based" or verbal modes of thinking with the activities of the two hemispheres of the brain. For lack of space, I cannot go into detail; in any case, I am more interested in the mechanisms of the synthesis of the two modes than reasons for their separation.

[4]$O(\ )$-NOTATION. We give a few words about $O(\ )$-notation for orders of magnitude of functions of natural argument n: we say that $f(n) = O(g(n))$ if there is a constant C such that $f(n) \leqslant Cg(n)$ for all sufficiently large n. Hence $f(n)$ is $O(n^d)$ if $f(n) \leqslant Cn^d$.

[5]KOBLITZ'S THESIS. Like all general proclamations about mathematics, Koblitz's thesis has its natural limits of applicability. As usual, we have all possible complications caused by the non-constructive nature of many mathematical proofs: sometimes it is possible to prove that a certain algorithm has polynomial complexity $O(n^d)$, without having any way to find the actual degree d; one example can be found in [330]. But we do not venture into this exciting, but dangerous, territory.

[6]CHOICELESS ALGORITHMS. A few words of warning are due. Any algorithm on a clearly described finite set of inputs can be made into a choiceless algorithm by running it on all possible reorderings of the input structures. Therefore the concept of choiceless computing is meaningful only if we assume resource limitations and focus on choiceless *polynomial time* algorithms.

[7]We give an interesting comment from Chris Hobbs on the formula

$$x_{1,2} = \frac{-b \pm \sqrt{b^2 - 4ac}}{2a}$$

for the roots of the quadratic equation:

> I know that this formula is always written with the plus/minus but, as you've argued in the text, it's not only unnecessary, it's also wrong. It has worried me since I first met it as a child that, according to that formula, there are four roots to a quadratic equation: the square root function delivers two and the plus/minus turns them into four (two positive, ++ and −−, and two negative, −+ and +−). Just a quibble but it's a shame that we don't have a notation for "the negative square root of x" and "the positive square root of x".

[8]TROPICAL MATHEMATICS. The works [396, 408] use min, not max, as a basic operation $\oplus$, but this makes no difference since in both cases the results are similar. Traditionally, max is used in works originating in control theory and min in papers motivated by applications to algebraic geometry.

[9]SUDOKU. Sudoku enthusiasts would not forgive me if I move from the subject without giving a single Sudoku puzzle. Here is one kindly provided by Gordon Royle, University of Western Australia. It contains only 17 clues (filled squares), but is still *deterministic*, that is, can be filled in, in only one way. Apparently the existence of deterministic Sudoku puzzles with less than 17 clues is an open problem.

							1	
4								
	2							
				5		4		7
		8				3		
		1		9				
3			4			2		
	5		1					
			8		6			

2

What You See Is What You Get

2.1 The starting point: mirrors and reflections

I use, as one of the principal running themes of this book, a comparison between two approaches to the concept of symmetry as it is understood and used in modern algebra. The corresponding mathematical discipline is well established and is called the *theory of finite reflection groups*. The reader should not worry if he or she has never encountered this name; as will soon be seen, the subject has many elementary facets.

To start with, the principal objects of the theory can be defined in the most intuitive way. First I give an informal description:

> Imagine a few (semi-transparent) mirrors in ordinary three-dimensional space. Mirrors (more precisely, their images) multiply by reflecting in each other, as in a kaleidoscope or a gallery of mirrors. Of special interest are mirror systems of which generate only finitely many reflected images. Such finite systems of mirrors happen to be one of the cornerstones of modern mathematics and lie at the heart of many mathematical theories.

As usual, the full theory is concerned with the more general case of n-dimensional Euclidean space, with 2-dimensional mirrors replaced by $(n-1)$-dimensional *hyperplanes*. To that end, we give a formal definition:

> A system of hyperplanes (mirrors or images of mirrors) $\mathbb{M}$ in the Euclidean space $\mathbb{R}^n$ is called *closed* if, for any two mirrors M_1 and M_2 in $\mathbb{M}$, the mirror image of the mirror M_2 in the mirror M_1 also belongs to $\mathbb{M}$ (Figure 2.1).

Anna Borovik,
née Vvedenskaya,
aged 8

Thus, the principal objects of the theory are *finite closed systems of mirrors*. In more evocative terms, the theory can be described as the geometry of *multiple mirror images*. This approach to symmetry is well known and is found, for example, in Chapter 5, §3 of Bourbaki's classical text [323][1], or in Vinberg's paper [415]; I have recently used it in my textbook *Mirrors and Reflections* [294].

However, closed systems of mirrors are usually known in mathematics under a different name, and in a completely different dress, as *finite reflection groups*. They make up a classical chapter of mathematics, which originated in the seminal works of H. S. M. Coxeter [336, 337] (hence yet another name: *finite Coxeter groups*). The theory can be based on the concept of a group of transformations (as is done in many excellent books; see, for example, [360, 367]) and can be developed in group-theoretic terms.

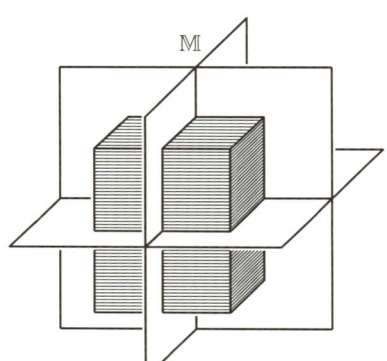

The system $\mathbb{M}$ of all mirrors of symmetry of a geometric body Δ is *closed*: the reflection of a mirror in another mirror is yet another mirror. Notice that if Δ is compact (i.e., closed and bounded), then all mirrors have a point in common.

Fig. 2.1. A closed system of mirrors. Drawing by Anna Borovik.

So we have two treatments, in two different mathematical languages, of the same mathematical theory (which I will call *Coxeter theory*). This is by no means an unusual thing in mathematics. What makes mirror systems/Coxeter groups interesting is that a closer look at the corresponding mathematical languages reveals their cognitive (and even neurophysiological!) aspects, much more obviously than in the rest of mathematics. In particular, as we shall soon see, the mirror system/Coxeter group alternative precisely matches the great *visual/verbal* divide of mathematical cognition.

It is worthwhile to pause for a second over the question of why we pay special attention to visual and speech processing. The answer is obvious: of all our senses, sight and hearing have the highest information processing rate and are used for communication. One can only speculate what mathematics would look like if we had an echolocating capacity (see a discussion of the way bats perceive the world in a paper by Kathleen Akins [146]). This is even more mind-boggling: try to imagine that humans have electric sensing and communicating facilities of the kind that Nile elephant fish *Gymnarchus niloticus* have and therefore that we live in a landscape made not of shapes and volumes (sight is of no use in the murky waters of the Nile) but of electromagnetic capacities and conductivities.[2] Would the concepts and results of vector calculus be self-evident to us? And a further question can be asked: which immediately intuitive mathematical concepts would become less intuitive? In Section 4.4 I attempt to suggest a partial answer to this question.

I wish to stress that, although the theory of Coxeter groups formally belongs to "higher" mathematics, the issues raised in the next two chapters are relevant to the teaching and understanding of mathematics at all levels, from elementary school to graduate studies. Indeed, I will be talking about such matters as *geometric intuition*. I will also touch on the role of pictorial proofs and self-explanatory diagrams; some of these may seem naïve, but, as I hope to demonstrate, they frequently lead deep into the heart of mathematics (see Section 2.6 for one of the more striking cases).

2.2 Image processing in humans

The mirror is one of the most powerful and evocative symbols of our culture; seeing oneself in a mirror is equated to self-awareness. But the reason why the language of mirrors and reflections happens to be so useful in the exposition of mathematical theories lies not so much at a cultural as at a psychophysiological level.

How do people recognize mirror images? Tarr and Pinker [235] showed that recognition of mirror images of planar shapes is done by subconscious mental rotation of 180° about an appropriately chosen axis. Remarkably, the brain computes the position of this axis!

This is how Pinker describes the effect of their simple experiment.

Erich Ellers,
aged 7

> So we showed ourselves [on a computer screen] the standard upright shape alternating with one of its mirror images, back and forth once a second. The perception of flipping was so obvious that we didn't bother to recruit volunteers to confirm it. When the shape alternated with its upright reflection, it seemed to pivot like

a washing machine agitator. When it alternated with its upside-down reflection, it did backflips. When it alternated with its sideways reflection, it swooped back and forth around the diagonal axis, and so on. *The brain finds the axis every time.* [218, pp. 282–283]

Interestingly, the brain does exactly the same with randomly positioned three-dimensional shapes, provided they *have the same chirality* (that is, both are of left-hand or right-hand type, as gloves, say) and can be identified by a rotation [233]. The interested reader may wish to take any computer graphics package which allows animation and see it for himself or herself.[3]

In view of these experiments, it is difficult to avoid the conclusion that Euler's classical theorem is hardwired into our brains:

> If an orientation-preserving isometry of the affine Euclidean space $\mathbb{AR}^3$ has a fixed point, then it is a rotation around some axis.

This illusion of rotation disappears when the brain faces the problem of the identification of three-dimensional mirror images of *opposite chirality*; indeed, they can still be identified by an appropriate rotation, but, this time, in four-dimensional space. The environment which directed the evolution of our brain never provided our ancestors with four-dimensional experiences.

It is difficult to avoid the conclusion that Euler's Theorem is hardwired into our brains.

Human vision is a solution of an ill-posed inverse problem of recovering information about three-dimensional objects from two-dimensional projections on the retinas of the eyes. Pinker stresses that this problem is solvable only because of the multitude of assumptions about the nature of the objects and the world in general built into the human brain or acquired from previous experiences.[4]

The algorithm of the identification of three-dimensional shapes is only one of many modules in the immensely complex system of visual processing in humans. It is likely that various modules are implemented as particular patterns of connections between neurons. It is natural to assume that different modules developed at different stages of evolution [234]. The older ones are likely to be simpler and involve relatively simple wiring diagrams. But since they had adaptive value, they were inherited and they acted as constraints in the evolution of later additions to the system, in particular any new modules which happened to process the outputs of, and interact with, the pre-existent modules. At every stage, evolution led to the development of an algorithm for solving a very special and narrow problem. Of course, the evolution is guided by the universal and basic principle of the survival of the fittest. But, translated into selection criteria for the gradual improvement of light-sensing organs, the general principle became highly specialized and changing over time. First it favored higher sensitivity to diffuse light of a few cells which previously had quite different functions; at the next step of evolution it favored individuals with light-sensitive cells positioned in a more efficient way and, most probably, only after that, started to favor individuals who had some mechanism for discriminating between light stimuli applied to different groups of cells. Therefore we should expect that the image processing algorithms of our brain have a multilayered structure which reflects their ontogenesis—and, not unlike many modern software systems, are full of outdated "legacy code".

The "flipping" algorithm for the recognition of mirror images of a flat object and the closely related (and possibly identical) "rotation" algorithm for making randomly orientated three-dimensional objects coincide provide rare cases where we can glimpse the inner workings of our mind. Observe, however, that the algorithms are solutions of relatively simple mathematical problems with a very rigid underlying mathematical structure, namely, the group of isometries of three-dimensional Euclidean space. There is no analogue of Euler's Theorem for four-dimensional space![5]

David Broomhead,
aged 8

The reader has possibly noticed that I prefer to use the term "algorithm" rather than "circuit", emphasizing the strong possibility that a given algorithm can be implemented by different circuit arrangements if some of the arrangements become impossible as the result of trauma, especially during the early stages of a child's development.

Studies of compensatory developments are abundant in the literature. When I was looking for some recent studies, my colleague David Broomhead directed me to the paper [182], a case study of a young woman who has been unable to make eye movements since birth but has surprisingly normal visual perception. This is astonishing because the so-called saccadic movements of the eyes are crucial for tracing the contours and the key features of objects. Try to experiment with a mirror: you will not see your eyes moving. During each saccade, the eye is in effect blind. We see the world frame-by-frame, as in the cinema. The continuity of the moving world is the result of the work of sophisticated interpolating routines integrated into the visual processing modules of our brain. Not surprisingly, continuity is one of the most intuitive (although hard to formalize) concepts of mathematics.

The woman in the study reported in [182] compensates for her lack of eye movement by quick movements of her head which follow the usual highly regular patterns of saccadic movements. I quote from the paper: "Her case suggests that saccadic movements, of the head or the eye, form the *optimal sampling method* for the brain". The italics are mine, since I find the choice of words very suggestive: mathematics is encroaching on the inner working of the brain, raising some really interesting metamathematical questions.

2.3 A small triumph of visualization: Coxeter's proof of Euler's Theorem

If you need convincing that visualization is a purposeful tool in learning, teaching, and doing mathematics, there is no better example than the proof of Euler's Theorem as given by Coxeter [338, p. 36]; I quote it *verbatim*. Remember that Coxeter's book was first published in 1948, so it was written for readers who were likely to have taken a standard course of Euclidean geometry and therefore had developed their geometric imagination.

> In three dimensions, a congruent transformation that leaves a point **O** invariant is the product of at most three reflections: one to bring together the two x-axes, another for the y-axes, and a third (if necessary) for the z-axes.
>
> Since the product of three reflections is opposite, a direct transformation with an invariant point **O** can only be the product of reflections in *two* planes through **O**, i.e., a rotation.

Erich Ellers,
aged 15

I add just a few comments to facilitate the translation into modern mathematical language: a *congruent transformation* is an isometry; a *direct transformation* preserves the orientation (chirality), while an *opposite* transformation changes it. Coxeter refers to the fact that the product of two mirror reflections is a rotation about the line of intersection of the mirrors. This is something that everyone has seen in a tri-fold dressing table mirror; the easiest way to prove the fact is to notice that the product of two reflections leaves invariant every point on the line of intersection of the mirrors.[6]

We humans are blessed with a remarkable piece of mathematical software for image processing directly hardwired into our brains. Coxeter made full use of it and expected the reader to use it, in his lightning proof of Euler's Theorem. (See a further discussion of Coxeter's proof in Section 6.3.) The perverse state of modern mathematics teaching is that "geometric intuition", the skill of solving geometric problems by looking at (simplified) two- and three-dimensional models, has been largely expelled from the classroom practice.[7]

However, our geometric intuition involves at least two quite different (although closely related) cognitive components: visual processing and motor control. The latter is paradoxical; our hands can move and act with extreme precision, but we receive much less information feedback from the feeling of the motion itself or from the position of our body and hands.

> *The perverse state of modern mathematics teaching is that "geometric intuition" has been largely expelled from classroom practice.*

To illustrate mathematical implications of this difference, I offer a small problem directly related to Euler's Theorem. I quote it from the book by David Henderson and Daina Taimina [272], where it is discussed in a slightly different context:

> When grinding a precision flat mirror, the following method is sometimes used: Take three approximately flat pieces of glass and put pumice between the first and second pieces and grind them together. Then do the same for the second and the third pieces and then for the third and first pieces. Repeat many times and all three pieces of glass will become very accurately flat.

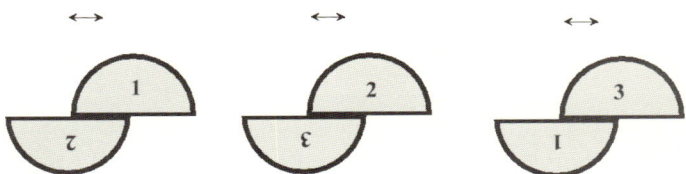

Fig. 2.2. Grinding a plane mirror (after David Henderson and Daina Taimina [272]).

See Figure 2.2. Now close your eyes and try to imagine your hands gently sliding one piece of glass all over the other. Do you see why this works?

Now I separate the question into two sub-questions, which, I believe, refer to two different levels of our intuition.

(A) Why do we need *three* pieces of glass to achieve perfect flatness? [?]

Indeed, why?

(B) Here is a trickier question: if only two pieces of glass are used, and the resulting surface is not plane, then (assuming that the grinding was thorough and even) what is this surface? [?]

Answer it!

The reader may wish to ponder these questions for a while; I give the answers in Section 4.5.

It is futile to talk about mathematical practice without first acknowledging that it can only be understood alongside its interaction with the human brain.

I am writing this book because I believe it is futile to talk about mathematical practice without first acknowledging that mathematics is an activity of the human mind and, in particular, the human brain. But our mind—or our cognitive system—is not homogeneous: its different parts developed at different stages of evolution, they have different levels of sophistication, an interaction between different modules is frequently awkward. We will not get much understanding of how mathematics lives in our minds without taking into account all the complexities and limitations of its constituent parts.

2.4 Mathematics: interiorization and reproduction

> *What is Mathematics, Really?*
> Reuben Hersh [48]
>
> *But Didactylos posed the famous philosophical conundrum:*
>
> *"Yes, But What's It **Really** All About, Then, When You Get Right Down To It, I **Mean** Really?"*
>
> Terry Pratchett [434, p. 167]

I have already quoted Davis and Hersh [21, p. 399], to say that mathematics is

> *the study of mental objects with reproducible properties.*

A famous mathematician, David Mumford, uses this formulation in his paper [72, p. 199] and further comments on it:

> I love this definition because it doesn't try to limit mathematics to what has been called mathematics in the past but really attempts to say why certain communications are

classified as math, others as science, others as art, others as gossip. Thus reproducible properties of the physical world are science whereas reproducible mental objects are math. Art lives on the mental plane (the real painting is not the set of dry pigments on the canvas nor is a symphony the sequence of sound waves that convey it to our ear) but, as the post-modernists insist, is reinterpreted in new contexts by each appreciator. As for gossip, which includes the vast majority of our thoughts, its essence is its relation to a unique local part of time and space.

If we accept this definition of mathematics, then we have to address two intertwined aspects of learning and mastering mathematics:

- the development of reproduction techniques for our own mental objects,
- interiorization of other people's mental objects.

There is a natural hierarchy of the methods of reproduction. A partial list in roughly descending order includes: proof; axiomatization; algorithm; symbolic and graphic expression. I wish to make it clear that reproduction is more than communication: you have to be able to reproduce your own mental work *for yourself*. Maybe it even makes sense to view *recovery procedures* for lost or forgotten mathematical facts as a distinct group of reproduction methods, as they have very specific features; see Chapter 9 for a more detailed discussion of recovery procedures.

Interiorization is less frequently discussed. For our purposes, we mention only that it includes visualization of abstract concepts; transformation of formal conventions into psychologically acceptable "rules of the game"; development of subconscious "parsing rules" for processing strings of symbols (most importantly, for reading mathematical expressions). At a more mundane level, one cannot learn an advanced technique of symbolic manipulation without first polishing one's skills in more routine computations to the level of almost automatic perfection. Interiorization is more than understanding; to handle mathematical objects effectively, one has to imprint at least some of their functions at the *subconscious* level of one's mind.

Interiorization is more than understanding; to handle mathematical objects effectively, one has to imprint at least some of their functions at the subconscious *level of one's mind.*

My use of the term "interiorization" is slightly different from the understanding of this word, say, by Weller et al. [142]. I put emphasis on the subconscious, neurophysiological components of the process. Meanwhile, I am happy to borrow from [142] the terms *en-capsulation* (and the reverse procedure, *de-encapsulation*) to stand for the conversion of a mathematical procedure, a learned sequence of action, into an object. The processes of encapsulation and de-encapsulation are one of the principal themes of the book; see Section 6.1 for a more detailed discussion.

It has to be clarified that reproduction does not mean *repetition*. It is a popular misconception that mathematics is a dull repetitive activity. Actually, mathematicians are easily bored by repetition. Perhaps this could create some difficulty in neurological studies of mathematics.

It is a popular misconception that mathematics is a dull repetitive activity.

Certain techniques for study of patterns of activation of the brain are easier to implement when the subject is engaged in an activity which is relatively simple and can be repeated again and again, so that the data can be averaged and errors of measurements suppressed. This works in studies like [203] which compared activation of the brains of amateur and professional musicians during actual or imagined performance of a short piece of violin music. Indeed, you can ask a musician to play the same several bars of music 10, 20, perhaps even 100 times—this is what they do in rehearsals. But it is impossible to repeat the same calculation 20 times: very soon the subject will remember the final and intermediate results. Moreover, most mathematicians will treat as an insult a request to repeat a similar calculation 20 times with varying data.

Some mathematical activities are of a compound nature and can be used as means of both interiorization and reproduction. A really remarkable one is the generation of examples, especially very simple (ideally, the simplest possible) examples—as

Proof is the key ingredient of the emotional side of mathematics.

discussed in Section 1.1. Really useful examples can be loosely divided into two groups: "typical", generic examples of the theory; or "simplest possible", almost degenerate examples, which emphasize the limitations and the logical structure of the theory. Of course, one of the attractive features of Coxeter Theory is that it is saturated by beautiful examples of both kinds; I discuss some "simplest" cases in Section 2.6.

Proof, being the highest level of reproduction activity, has an important interiorization aspect: as Yuri Manin stresses in his book *Provable and Unprovable*, a proof becomes such only after it is *accepted* (as the result of a highly rigorous process) [383, pp. 53–54]. Manin describes the act of acceptance as a social act; however, the importance of its personal, psychological component can hardly be overestimated. One also should note that proof is the key ingredient of the emotional side of mathematics; proof is the ultimate explanation of *why* something is true, and a good proof often has a powerful emotional impact, boosting confidence and encouraging further questions "why?".

Visualization is one of the most powerful techniques for interiorization. It anchors mathematical concepts and ideas firmly into one of the most powerful parts of our brain, the visual processing module. Returning to the principal example of this book, mirrors and reflections, I want to point out that finite reflection groups allow an approach to their study based on a systematic reduction of this whole range of complex geometric configurations to simple two- and three-dimensional special cases. Mathematically this is expressed by a theorem:

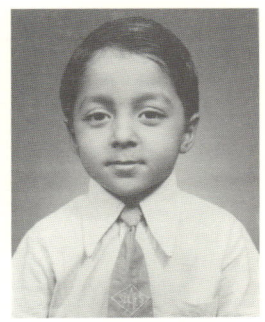

Satyan L. Devadoss,
aged 3.
© Satyan L.
Devadoss

a finite reflection group is a Coxeter group.

Avoiding a technical discussion, this means, in particular, that all relations between elements in a reflection group are consequences of relations between *pairs* of generating reflections. But a pair of mirrors in the n-dimensional Euclidean space is no more sophisticated a configuration than a pair of lines on the plane, and all the properties of the former can be deduced from that of the latter. *This provides a mathematical explanation of why visualization is such an effective tool in the theory of finite reflection groups.*

2.5 How to draw an icosahedron on a blackboard

My understanding of visualization as an interiorization technique leads me to believe that drawing pictures, and devising new kinds of pictures to draw, is an important way of facilitating mathematical work. This means that pictures have to be treated as mathematical objects and, consequently, must be *reproducible*. Students in the classroom should be able to *draw right away* the figures we put on the blackboard.

I have to emphasize the difference between *drawings* or *sketches* which are supposed to be reproduced by the reader or student and more technically sophisticated illustrative material (I will call these *illustrations*), especially computer-generated images designed for the visualization of complex mathematical objects (see a book by Bill Casselman [327] for an introduction into the art of illustrating mathematical texts). It would be foolish to impose any restrictions on the technical perfection of illustrations. However, one should be

Fig. 2.3. What different nations eat and drink. A statistical diagram from a calendar published in Austro-Hungaria in 1901. Source: Marija Dalbello [19], reproduced with permission. See [20] for a discussion of the historical context.

This style of graphical representation of quantitative information strikes us now as patronizing and non-mathematical. It would be interesting to trace the cultural change over the 20th century: why do we expect a much more slim and abstract mode of presentation of information? Is this a result of the visual information overload created by TV and the Internet? It is worth mentioning that the level of basic numeracy in the middle classes of the Austro-Hungarian Empire, the target readership of the *Šareni svjetski koledar*, was almost definitely higher than in modern society.

aware of the danger of excessive details; as William Thurston—one of the leading geometers of our time—stresses,

> words, logic and detailed pictures rattling around can inhibit intuition and associations. [94, p. 165]

For that reason I believe that *drawings* should be intentionally very simple, even primitive. Mathematical pictures represent *mental* objects, not the real world! In the words of William Thurston,

> [people] do not have a very good built-in facility for *inverse vision*, that is, turning an internal spatial understanding back into a two-dimensional image. Consequently, mathe-

maticians usually have fewer and poorer figures in their
papers and books than in their heads. [94, p. 164]

We have to be careful with our drawings and make sure that they
correctly represent our "internal spatial understanding".

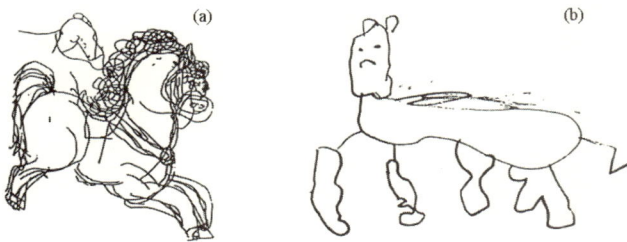

Fig. 2.4. Vision vs. "inverse vision": (a) a picture by Nadia (Drawing
3 from Selfe [228], © 1977 Elsevier, reproduced with permission) as
opposed to (b) a picture by a normal child (Snyder and Mitchell [231],
reproduced with permission). See [231] for a detailed discussion.

The pictures in Figure 2.4, taken from Selfe [228] and Snyder
and Mitchell [231], illustrate the concept of "inverse vision" as introduced by Thurston. The picture (a) on the left is drawn from
memory by Nadia, a three-and-a-half year old autistic child who at
the time of making the picture has not yet developed speech [228].
Picture (b) is a representative drawing of a normal child, at age
four years and two months. It is obvious that a normal child draws
not a horse, but a concept of a horse.

Fig. 2.5. Horses by Nadia, at age of 3 years 5 months (Drawing 13 from
Selfe [228], © 1977 Elsevier, reproduced with permission).

Fig. 2.6. Horses from Chauvet Cave (Ardeche). Document elaborated with the support of the French Ministry of Culture and Communication, Regional Direction for Cultural Affairs—Rhône-Alpes, Regional Department of Archaeology.

Nicholas Humphrey [195] drew even bolder conclusions from Nadia's miraculous drawings. He observed that Nadia's pictures have a most suggestive resemblance to cave paintings of 30,000–20,000 years ago—compare Figures 2.5 and 2.6. Humphrey conjectured that human language developed in two stages. At the first stage it referred only to people and relations between people; the natural world (including animals) had no symbolic representations in the language and therefore early people had no symbols for the external world. Cave paintings such as the one in Figure 2.6, are symbolically unprocessed images on the retina of the painter's eye, placed one over another without much coordination or a coherent plan. At the same time, people could already have words and symbols which referred to other people—which is consistent with the simultaneous presence, in some cave paintings, of strikingly realistic animals and highly schematic human figures; see Figure 2.7.

Mathematical pictures are symbolic images, not representations of reality.

Mathematical pictures are symbolic images, not representations of reality. Like a matchstick human in Figure 2.7, they are produced by "inverse vision". I dare to say that they do not belong to art. I propose that image processing which leads to the creation of paintings and drawings in the visual arts is different from that of mathematics.[8] Mathematical pictures therefore should not provoke an inferiority complex in readers who have not tried to draw anything since their days in ele-

Fig. 2.7. A symbolic human and a naturalistic bull. Rock painting of a hunting scene, c. 17000 BC/ Caves of Lascaux, Dordogne, France. Source: *Wikipedia Commons*. Public domain.

mentary school; they should instead act as an invitation to readers to express their own mental images.

Figure 2.8 illustrates the most effective way of drawing an icosahedron, so simple that it is accessible to the reader with very modest drawing skills. First we mark symmetrically positioned segments in an alternating fashion on the faces of the cube (left) and then connect the endpoints (right). The drawing actually provides a proof of the existence of the icosahedron: varying the lengths of segments on the left cube, it is easy to see from continuity principles that, at a certain length of the segments, all edges of the inscribed polyhedron on the right become equal. [?] Moreover, this construction helps to prove that the group of symmetries of the resulting icosahedron is as big as one would expect it to be; see [294] for more details.

In a unit cube, find the length of the segments which makes all triangle faces equilateral; Figure 2.8.

Figure 2.8 works as a proof because it is produced by "inverse vision". To draw it, you have to run, in your head, the procedure for the construction of the icosahedron. And, of course, the continuity principles used are self-evident—they are part of the same mechanisms of perception of motion which glue, in our minds, the cinema's 24 frames per second into continuous motion.

I hope that now you will agree that Figure 2.8 deserves to be treated as a mathematical statement. It is useful to place it in a wider context. Notice that construction of the icosahedron is the same thing as construction of the finite reflection group H_3; this can be done by means of linear algebra—which leads to rather nasty calculations—or by means of representation theory—which

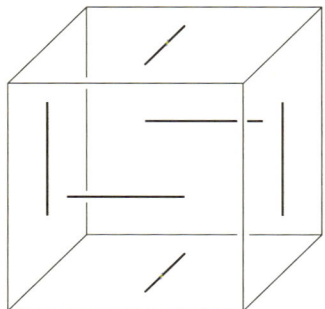 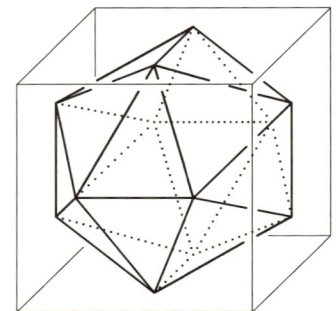

Fig. 2.8. A self-evident construction of an icosahedron. Drawing by Anna Borovik.

This construction of the icosahedron is adapted from the method of H. M. Taylor [47, pp. 491–492]. John Stillwell has kindly pointed out that it goes back to Piero della Francesca and can be found in his unpublished manuscript *Libellus de quinque corporibus regolaribus* from around 1480.

requires some knowledge of representation theory. It also can be done by quaternions—which is nice and beautiful but requires knowledge of quaternions. The graphical construction is the simplest; using computer jargon, it is a WYSIWYG ("What You See Is What You Get") mode of doing mathematics, which deserves to be used at every opportunity.

2.6 Self-explanatory diagrams

This section is more technical and can be skipped.

Self-explanatory diagrams have been virtually expunged from modern mathematics. I believe they can be useful, not only in proofs, etc., but also as the means of a metamathematical discussion of the structure and interrelations of mathematical theories.

Figure 2.9 is one example, taken from *Mirrors and Reflections* [294]: the isomorphism of the root systems D_3 (shown on the left, inscribed into the unit cube $[-1, 1]^3$) and A_3 is not immediately obvious, but the corresponding mirror systems coincide most obviously. The mirror system D_3 (the system of mirrors of symmetry of the cube) is shown in the middle by tracing the intersections of mirrors with the surface of the cube and, on the right, by intersections with the surface of the tetrahedron inscribed in the cube. Comparing the last two pictures, we see that the mirror system of type D_3 is isomorphic to the mirror system of the regular tetrahedron, that is, to the system of type A_3.

As we shall soon see, this isomorphism has far-reaching implications.

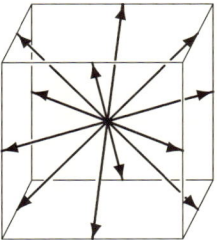 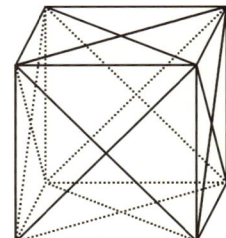 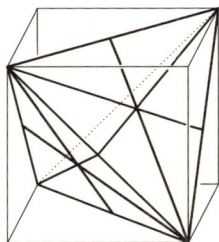

Fig. 2.9. An example of a self-explanatory diagram. Drawing by Anna
Borovik.

Indeed, at the level of complex Lie groups the isomorphism
$D_3 \simeq A_3$ becomes a rather mysterious isomorphism between the
six-dimensional orthogonal group $\mathrm{SO}_6(\mathbb{C})$ and $\frac{1}{2}\mathrm{SL}_4(\mathbb{C})$, the factor
group of the four-dimensional special linear group $\mathrm{SL}_4(\mathbb{C})$ by the
group of scalar matrices with diagonal entries ± 1 (or, if you prefer
to work with spinor groups, between $\mathrm{Spin}_6(\mathbb{C})$ and $\mathrm{SL}_4(\mathbb{C})$).

This is not yet the end of the story. The compact form of $\mathrm{SL}_4(\mathbb{C})$
is SU_4, and hence the embedding

$$\mathrm{SU}_4 \hookrightarrow \mathrm{Spin}_6(\mathbb{C})$$

features prominently in the representation theory of SU_4, and
hence in the SU_4-symmetry formalism of theoretical physics.

But the underlying reason for the isomorphisms retains all the
audacity of Keplerian reductionism: the tetrahedron can be inscribed into the cube. Compare with Figure 2.10.

Because of their truly fundamental role in mathematics, even
the simplest diagrams concerning finite reflection groups (or finite
mirror systems, or root systems—the languages are equivalent)
have interpretations of cosmological proportions. Figure 2.11 is
even more instructive. It is a classical case of the *simplest possible
example* as discussed in Chapter 1. For example, it is the simplest
rank 2 root system, or the simplest root system with a non-trivial
graph automorphism; the latter, as we shall see in a minute, has
really significant implications.

Figure 2.11 also demonstrates that the root system $D_2 =
\{\,\pm\epsilon_1 \pm \epsilon_2\,\}$ is isomorphic to $A_1 \oplus A_1 = \{\,\pm\epsilon_1, \pm\epsilon_2\,\}$. At the level
of Lie groups, this isomorphism plays an important role in the de-
scription of the structure of four-dimensional space-time of special
relativity; namely, it yields the structure of the Minkowski group
(the group of isometries of the four-dimensional space-time of spe-

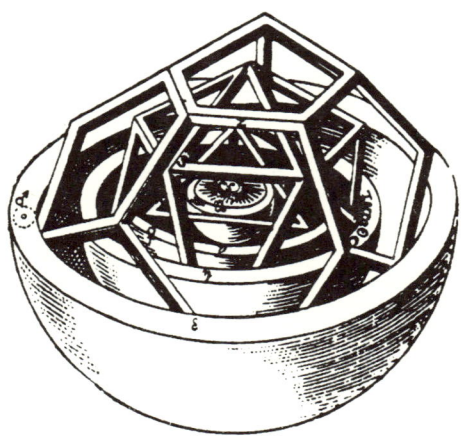

Fig. 2.10. A fragment of a famous engraving from Kepler's *Mysterium Cosmographicum*. Public domain.

cial relativity theory with the metric given by the quadratic form $x^2 + y^2 + z^2 - t^2$).

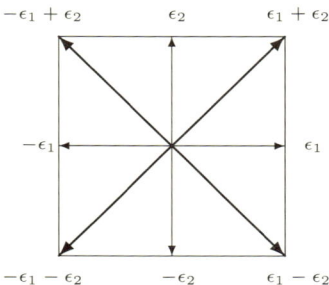

Fig. 2.11. This diagram demonstrates the isomorphism of the root systems $D_2 = \{ \pm\epsilon_1 \pm \epsilon_2 \}$ and $A_1 \oplus A_1 = \{ \pm\epsilon_1, \pm\epsilon_2 \}$. Drawing by Anna Borovik.

Indeed, the isomorphism of root systems $D_2 \simeq A_1 \oplus A_1$ leads to the isomorphisms

$$\mathrm{Spin}_4(\mathbb{C}) \simeq \mathrm{SL}_2(\mathbb{C}) \times \mathrm{SL}_2(\mathbb{C})$$

and

$$\mathrm{SO}_4(\mathbb{C}) \simeq \mathrm{SL}_2(\mathbb{C}) \otimes \mathrm{SL}_2(\mathbb{C})$$

(the tensor product of two copies of $SL_2(\mathbb{C})$, each acting on its canonical two-dimensional space $\mathbb{C}^2$). The connected component of the Minkowski group is a real form of $SO_4(\mathbb{C})$. Hence it is the group of fixed points of some involutory automorphism τ of the group $SO_4(\mathbb{C})$. What is this automorphism τ? Let us look again at the quadratic form $x^2 + y^2 + z^2 - t^2$; it is a real form of the complex quadratic form $z_1^2 + z_2^2 + z_3^2 + z_4^2$ but has lost the symmetric pattern of coefficients. One can see that this means that τ swaps the two copies of $SL_2(\mathbb{C})$ in $SL_2(\mathbb{C}) \otimes SL_2(\mathbb{C})$ and therefore has to be the symmetry between the two diagonals of the square in Figure 2.11. Being an involution, τ fixes pointwise the "diagonal" subgroup in $SL_2(\mathbb{C}) \otimes SL_2(\mathbb{C})$ isomorphic to $PSL_2(\mathbb{C})$. (It is $PSL_2(\mathbb{C})$ rather than $SL_2(\mathbb{C})$ because its center $\langle -\text{Id} \otimes -\text{Id} \rangle$ is killed in the tensor product.) Hence the connected component of the Minkowski group is isomorphic to $PSL_2(\mathbb{C})$.

Three cheers for Kepler!

Notes

[1] *Groupes et Algebras de Lie, Chap 4, 5, et 6* is one of the better books by Bourbaki; it even contains a drawing, in an unexpected deviation from his usual aesthetics. See an instructive discussion of the history of this volume by its main contributor, Pierre Cartier [87].

[2] See discussion of electromagnetic imaging in fish in Nelson [209], Rasnow and Bower [221].

[3] To reproduce Tarr's experiments, I was using PAINTSHOP PRO, with three-dimensional images produced by XARA, two software packages picked up from the cover CD of a computer magazine.

[4] PICTORIAL PROOFS. Jody Azzouni [5, p. 125] commented on pictorial proofs that they work only because we impose many assumptions on diagrams admissible as part of such proofs. As he put it:

> We can conveniently stipulate the properties of *circles* and take them as mechanically recognizable because there are no *ellipses* (for example) in the system. Introduce (arbitrary) *ellipses* and it becomes impossible to tell whether what we have drawn in front of us is a *circle* or an *ellipse*.

It is likely that his remark would not surprise cognitive psychologists; they believe that this is what our brains are doing anyway.

[5] FOUR-DIMENSIONAL INTUITION. Here is an interesting question: can one be habituated in a four-dimensional space (say, with a flight simulator). Of course, a three-dimensional image, stereo or holographic, could help. To put the point more radically, do we learn the number of dimensions?

[6] EULER'S THEOREM. I accept that the reader has every right to insist, if so inclined, that the "best" way to prove Euler's Theorem is by reduction to algebra: the characteristic polynomial of a "generic" three-dimensional orthogonal matrix is a cubic with real coefficients, hence has a real root and a pair of conjugate complex roots; the orthogonality means that the eigenvalues have magnitude 1, hence should be equal to ± 1 and $\cos\theta \pm i\sin\theta$.

If the matrix has determinant $+1$, then the real eigenvalue is $+1$, and the corresponding eigenvector gives the direction of the axis of rotation, while θ is the angle of rotation. But is that really better than Coxeter's proof?

[7]WHY WAS GEOMETRIC INTUITION EXPELLED FROM CLASSROOM? One of the reasons is the predominance of written examinations. There are no intrinsic pedagogical reasons why oral examinations are inferior to written examinations: written assessment dominates the modern teaching because it creates a convenient audit trail. In a written examination, purely "algebraic" solutions are easier to write down and easier to mark. In an oral exam, a candidate's awkward sketch on a blackboard may be worth a thousand words.

As a result of the degradation of academic practice under the pressure of the audit culture, I have seen courses in linear programming taught without any reference to the geometric interpretation of its principal concepts.

[8]The situation could be different in ornamental art, especially when images of other people and of the natural world are prohibited by cultural conventions or religion. The creators of the Islamic mosaics in the Alhambra had in fact discovered most of the planar crystallographic groups—an intellectual achievement which firmly places their work in the realm of mathematics. See Branko Grünbaum [42] for one of the most up-to-date discussion of symmetry groups present in the Alhambra. Also, Lu and Steinhardt [66] discuss an even more striking discovery of medieval Islamic architecture: quasi-periodic tilings.

3

The Wing of the Hummingbird

3.1 Parsing

So far I have emphasized the role of visualization in mathematics and its power of persuasion. Here I will try to unite the visual and symbolic aspects of mathematics and touch upon the limitations of visualization.

Indeed, visualization works perfectly well in the geometric theory of *finite* reflection groups, but it needs to be refined for the more general theory of infinite Coxeter groups. We take a brief look at this more general theory, which is of special interest to us at this point. As truly fundamental mathematical objects, Coxeter groups provide an example of a theory where the links between mathematical teaching and learning and cognitive psychology lie exposed. Besides the power of geometric interpretation and visualization, the theory of Coxeter groups relies on manipulation of *words* in canonical generators (chains of consecutive reflections, in the case of reflection groups) and provides one of the best examples of the effectiveness of the *language metaphor* in mathematics.

It is tempting to try to link the psychology of symbolic manipulation in mathematics with the Chomskian conjecture that humans have an innate facility for parsing human language. Basically, parsing is the recognition/identification of the structure of a string of symbols (phonemes, letters, etc.). We parse everything we read or hear. Here is an example from Steven Pinker's book [219, pp. 203–205] where this thesis is vigorously promoted:

> Remarkable is the rapidity of the motion of the wing of the hummingbird.

To make sense of the phrase, we have to mentally bracket subphrases, resulting in something like the following:

[Remarkable is
    [the rapidity of
        [the motion of
            [the wing of
                [the hummingbird]]]]].

A sentence might have a different bracket pattern. Just compare

[Remarkable is [the rapidity of [the motion]]]

and

[[The rapidity [that [the motion] has]] is remarkable].

Some patterns are harder to deal with than others: for example,

[[The rapidity that [the motion that [the wing] has] has] is remarkable].

Some bracketings are close to incomprehensible, even though the sentence conveys the same message:

[[The rapidity that [the motion that [the wing that [the hummingbird] has] has] has] is remarkable]. [?] [1]

Gregory Cherlin kindly offered a brainteaser from his childhood: Punctuate: Smith where Jones had had had had had had had had had had the professor's approval.

Different human languages have different grammars, resulting in different parsing patterns. The grammar is not innate; Pinker emphasizes that what is innate is the human capacity to generate parsing rules. The generation of parsing patterns is a part of language learning (and young children are extremely efficient at it). It is also a part of the interiorization of mental objects of mathematics, especially when these *objects* are represented by strings of *symbols*.[2]

Cognitive scientists are very much attracted to case studies of "savants", autistic persons with an ability to handle arithmetic or calendrical calculations disproportionate to their low general IQ. As Snyder and Mitchell formulated it [231],

> . . . savant skills for integer arithmetic . . . arise from an ability to access some mental process which is common to us all, but which is not readily accessible to normal individuals.

The parsing mechanisms of the human brain are the key to the understanding of low-level arithmetic and formula processing.

What are these "hidden" processes? In one of the extreme cases (mentioned by Butterworth [160]), a severely autistic young man was unable to understand speech, but he could handle factors and primes in numbers. This suggests that certain mathematical actions are related not so much to language itself, but to the parsing facility, one of the components of the language system. An autistic person may have difficulty in handling language for reasons unrelated to his parsing ability; for example he may fail to recognize the source of speech communication as another person (or to understand the difference between what he knows and what the other person knows). But, in order to achieve such feats as "doubling 8 388 628 up to 24 times to obtain 140 737 488 355 328 in several seconds" [231, p. 589], an autistic person still has to be able to input into his brain the numbers given, inevitably, as strings of phonemes or digits.

I propose a conjecture that the parsing mechanisms of the human brain are the key to the understanding of low-level arithmetic and formula processing.

David Pierce,
aged 6

Moving several levels up the hierarchy of mathematical processes, we have a fascinating idea in the theory of automatic theorem proving: *rippling*, a formalization of a common way of mathematical reasoning where "formulae are manipulated in a way that increases their similarities by incrementally decreasing their differences" [325, p. 13]. This is facilitated by subdividing the formula into parts which have to be preserved and parts which have to be changed. Again, we see that in order to understand how humans use rippling in mathematical thinking (and whether they actually use it), we have to understand how our brains parse mathematical formulae.

To be on the cautious side, I am prepared to accept that parsing might be much more prominent in the input/output functions of the brain than in the internal processing of information. In a rare case of a savant with higher than normal general intellectual abilities, Daniel Tammet is able to vividly describe the way he perceives the world, language, and numbers. It is obvious from his words that number processing happens to be directly wired into the visual module of his brain. For him, many numbers have a unique visual form.

> "Different numbers have different colours, shapes and textures ... [The number] one is very bright and shining, like someone flashing a light into my face. Two is like a movement from right to left. Five is a clap of thunder or the sound of a wave against a rock. Six I find more difficult: it's more like a hole or a chasm. When I multiply numbers, I see two shapes in a landscape. The space between the images makes a third shape, like a jigsaw piece. And that third shape gradually crystallises: I see a fuzziness that becomes clearer and clearer." [430]

He adds that the whole process takes place in a flash, "like sparks flying off".

Although Daniel Tammet suffers from Asperger's syndrome (a form of autism) which to some degree inhibits his social skills—he has to remind himself that other people have thoughts entirely separate from his own and not to assume that they automatically know everything he knows—he has outstanding linguistic skills, speaks seven languages, and learned Icelandic in a week. He can also recite π to 22,514 decimal places. His case appears to confirm the thesis by Snyder and Mitchell; indeed, he has "an ability to access some mental process ... which is not readily accessible to normal individuals". This very access, however, requires parsing of the input.

3.2 Number sense and grammar

I turn to another remarkable insight from cognitive psychology, which links mechanisms of language processing to mastering arithmetic.

When infants learn to speak (in English) and count, there is a distinctive period, lasting five to six months, in their development, when they know the words *one, two, three, four* but can correctly apply only the numeral "one", when talking about a single object; they apply the words "two, three, four", apparently at random, to any collection of more than one object. Susan Carey [162] calls the children at this stage *one-knowers*. The most natural explanation is that they react to the formal grammatical structures of the adults' speech: *one doll*, but *two doll**S**, three doll**S***. At the next stage of development, they suddenly start using the numerals *two, three, four, five* correctly. Chinese and Japanese children become one-knowers a few months later—because the grammar of their languages has no specific markers for singular or plural in nouns, verbs, and adjectives.

In learning basic arithmetic, grammar precedes the words!

When the native language is Russian, the "one-knower" stage is replaced by the "one-(two-three-four) knower" stage, where children differentiate between three categories of quantities: single object sets, the sets of two, three, or four objects (without further differentiation between, say, two or three objects), and sets with five or more objects. This happens because morphological differentiation of plural forms goes further in Russian than in English.

When I heard about special plural forms of two, three, or four nouns in a lecture by Susan Carey at the *Mathematical Knowl-*

edge 2004 conference in Cambridge, I was mildly amused because it made no sense to me, as a native Russian speaker. Still, I started to write on note paper:

one doll	одна кукл**А**
two doll**S**	две кукл**Ы**
three doll**S**	три кукл**Ы**
four doll**S**	четыре кукл**Ы**
five doll**S**	пять кук**ОЛ**
⋮	⋮
ten doll**S**	десять кук**ОЛ**

I was startled: yes, Susan Carey was right! I had been using, all my life, the morphological rules for forming plurals—but using them subconsciously, without ever paying attention to them. But, apparently, an infant's brain is tuned exactly to picking up the rules: it is easier for the child to associate the number of objects with the morphological marker in the noun signifying the object than with the words *one* or *two*. The interested reader will find a detailed discussion of plurality marking in Sarnecka et al. [225]. Meanwhile, one of the readers of my blog brought my attention to an even more striking example: in Russian, in some rare cases, the whole noun changes, not just the plurality marker. For example, one year, two year**S**, three, four, five year**S** are translated into Russian as один год, два, три, четыре год**А**, пять **ЛЕТ**. Notice the same thresholds: one/two, four/five. *In learning basic arithmetic, grammar precedes the words!*

We shall return to the discussion of the four/five threshold in the context of subitizing and short-term memory, in Section 4.1. However, in the particular case of the Russian language, there is the possibility of a historic explanation for the peculiar behavior of plurality markers: they are remnants from the times when an Indo-European predecessor of the Russian language used a system of numerals based on the number 4 [438]. It might happen, however, that the historic explanation is only intermediate, since it does not answer the crucial question of why a base 4 system had appeared in the first instance and why, apparently, its subsequent evolution led to a bifurcation into a base 9 system of numerals (now extinct, but still traceable in formulae from Russian fairy tales: в тридевятом царстве, *in a three times ninth kingdom*) and the decimal one, now predominant.[3]

Barbara Sarnecka, aged 3

3.3 What about music?

It would be interesting to see to what extent the parsing mechanisms of language processing are at work in wider auditory perception; for example, are they relevant for the perception of music? Do
we parse notes by the same neurological mechanism which we use
for parsing phonemes? Unfortunately, I cannot regard myself as an
expert in music and therefore restrict myself to a few quotations.

My first quotation comes from a review, written by composer
Dorothy Kerr, of the recent book *Music and Mathematics* [32]
(strongly recommended!). Kerr, in effect, links music with the predictive nature of auditory processing.

> For a composer, some of the moments of greatest excitement
> lie in achieving a successful integration of 'mathematical'
> and 'musical' processes, though we may not think about it
> in these terms. Take the canon (a musical device that is es-
> sentially a translational symmetry) as an example: a very
> simple experiment that anyone can do is to set up a time
> delay between two copies of the same sound source (such
> as that produced when listening to digital radio simulta-
> neously with an analogue receiver).[4] At first—provided the
> time interval allows it to be readily perceived—this simple
> geometrical effect can be very engaging to the ear (given
> how easy it is to create a satisfying effect in this way it is
> perhaps not surprising that canon is one of the earliest and
> most prevalent devices of musical composition). After the
> canon we have made has been going for a while, the novelty
> wears off and we develop the need for some kind of change
> or a new layer of interest. The nature and precise timing of
> such alterations, a calculation we usually make using our
> intuition, is one of the most basic aspects of the art of com-
> position. [. . .] A process that is too obvious trails far behind
> the listener's ability to predict its outcomes. (Such music—
> to borrow the words of Harrison Birtwistle—'finishes before
> it stops'.)

The second quotation is from Thomas Mann's *Der Zauberberg*,
a book famous for—among other things—a detailed study of the
phenomenology of time. It describes music as parsing in its purest
form:

> "I am far from being particularly musical, and then the
> pieces they play are not exactly elevating, neither classic
> nor modern, but just the ordinary band-music. Still, it is a
> pleasant change. It takes up a couple of hours very decently;
> I mean it breaks them up and fills them in, so there is some-
> thing to them, by comparison with the other days, hours

and weeks that whisk by like nothing at all. You see an unpretentious concert-number lasts perhaps seven minutes, and those seven minutes amount to something; they have a beginning and an end, they stand out, they don't easily slip into the regular humdrum round and get lost. Besides they are again divided up by the figures of the piece that is being played, and these again into beats, so there is always something going on, and every moment has a certain meaning, something you can take hold of ... " (Translation by H. T. Lowe-Porter)

Of course, this gives only one dimension of music, essentially ignoring the harmony. In the words of Daniel Barenboim,

The music can only be of interest if the different strands of the polyphonic texture are played so distinctly that they can all be heard and create a three-dimensional effect—just as in painting, where something is moved into the foreground and something else into background, making one appear closer to the viewer than the other, although the painting is flat and one-dimensional.

I would not dare to venture further and I leave it to someone else to develop this wonderful theme.

3.4 Palindromes and mirrors

To illustrate the role of parsing and other word processing mechanisms in doing mathematics, let us briefly describe Coxeter groups in terms of words.

We work with an alphabet $\mathbb{A}$ consisting of finitely many letters, which we denote a, b, etc. A *word* is any finite sequence of letters, possibly empty (we denote the empty word ϵ). Notice that we have infinitely many words. To impose an algebraic structure onto the amorphous mass of words, we proclaim that some of them are equivalent to (or synonymous with) other words; we shall denote the equivalence of words V and W by writing $V \equiv W$. We demand that concatenation of words preserve equivalence: if $U \equiv V$, then $UW \equiv VW$ and $WU \equiv WV$: if *mail* is the same as *post*, then *mailroom* is the same as *postroom*. We denote the language defined by the equivalence relation $\equiv$ by $\mathcal{L}_{\equiv}$.

So far all that was just the proverbial "general nonsense" which we frequently find in the formal exposition of mathematical theories. Mathematicians treat such formalities with great respect but frequently ignore them in actual work; formal definitions play the same role as fine print in insurance policies. Beware the fine print when you make a claim!

It is remarkable how little we have to add in order to create the extremely rigid, crystalline structure of a Coxeter group. To that end, we say that a word is *reduced* if it is not equivalent to any shorter word. Now we introduce just two axioms which define *Coxeter languages*:

DELETION PROPERTY. If a word is *not* reduced, then it is equivalent to a word obtained from it by deleting some *two* letters.
(Of course, it may happen that the new word is still not reduced, in which case the process continues in the same fashion, two letters at a time.)

REFLEXIVITY. Words like *aa* obtained by doubling a letter are not reduced (hence are equivalent to the empty word, by the Deletion Property); *aardvark* is not a reduced word.

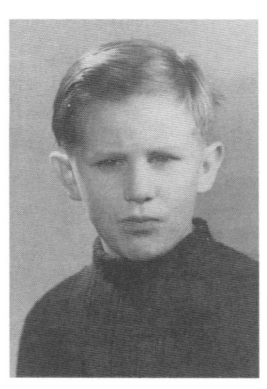

Michel Las Vergnas,
aged 9

Actually, a Coxeter language is exactly a Coxeter group, but I intentionally ignore this crucial (for a mathematician) fact and formulate everything in terms of words and languages.

I will now give a (straightforward) reformulation of a classical theorem of 20th-century algebra, due to Coxeter and Tits. My formulation is a bit of a caricature devised specifically for the purposes of the present book.

To emphasize the language aspects, let us make *palindromes*, that is, non-empty reduced words such as "level" that read the same backwards as forwards, the central object of the theory. [5]

Now the Coxeter–Tits Theorem becomes a theorem about *representation of palindromes by mirrors*.

The Palindrome Representation Theorem. Assume that a Coxeter language $\mathcal{L}_{\equiv}$ contains, up to equivalence, only finitely many palindromes.[6] Then:

- There exists a finite closed system $\mathbb{M}$ of mirrors in a finite-dimensional Euclidean space $\mathbb{R}^n$ such that the mirrors in $\mathbb{M}$ are in one-to-one correspondence with the equivalence classes of the palindromes.
- Moreover, if M_1 and M_2 are mirrors and P_1, P_2 their palindromes, then the palindrome associated with the reflected image of the mirror M_1 in the mirror M_2 is $P_2 P_1 P_2$, if the latter is reduced, or a palindrome obtained from the word $P_2 P_1 P_2$ by reduction.
- Finally, every closed finite system of mirrors in the Euclidean space $\mathbb{R}^n$ can be obtained in this way from the system of palindromes in an appropriate Coxeter language.

The interested reader will find the ingredients of a proof of this result in Chapters 5 and 7 of [320]. It involves, at some point, the

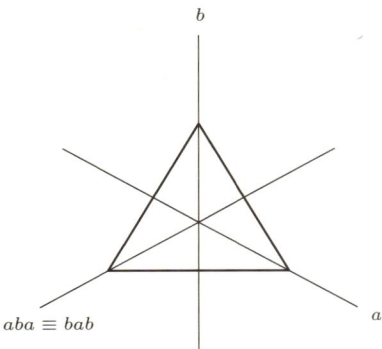

Fig. 3.1. The Palindrome Representation Theorem: The three mirrors of symmetry of the equilateral triangle correspond to the palindromes a, b, and aba. Together with the equivalences $aa \equiv bb \equiv \epsilon$ (the empty word), the equivalence $aba \equiv bab$ warrants that the corresponding Coxeter language does not contain any other palindromes.

following equivalence [294, Exercise 11.8]:

$$a_1 \cdots a_l \equiv a_l^{a_{l-1}\cdots a_1} \cdot a_{l-1}^{a_{l-2}\cdots a_1} \cdots a_2^{a_1} \cdot a_1,$$

where $b^{a_k \cdots a_1}$ is an abbreviation of a palindromic word

$$a_1 \cdots a_k \cdot b \cdot a_k \cdots a_1;$$

the foregoing identity expresses an arbitrary word as the concatenation of palindromic words; its proof consists of the rearrangement of brackets and the cancelation of doubled letters $a_i a_i$ whenever they appear. Proofs like that are one of the many reasons why, in order to master the theory of Coxeter groups expressed in a "linguistic" manner, the novice reader has to develop an ability to manipulate imaginary mental brackets with a rapidity comparable with only the remarkable rapidity of the motion of the wing of the hummingbird.

I reiterate that I devised the palindrome formulation of the Representation Theorem specifically for the needs of the present book. When afterwards I made a standard search on GOOGLE and MathSciNet [446], I was pleased to discover that my formulation appeared to be new.

We can reuse space, but, unfortunately, we cannot reuse time.

I was also pleasantly surprised to find more than a hundred papers on palindromes produced by computer scientists. The set of all palindromic words in a given alphabet is one of the simplest examples of a language which can be generated only by a device with some kind of memory, say, with a stack or push-down storage which works on the principle "last in—first out", like bullets in a handgun clip. It makes palindromes a very attractive test problem in the study of the complexity of word processing, for example, for comparing two fundamental concepts of algorithmic complexity: space-complexity, measured by the amount of memory required, and time-complexity. The difference between the two complexities is deeply philosophical: we can reuse space, but, unfortunately, we cannot reuse time. I was particularly fascinated to learn that palindromes are recognizable by Turing machines working within sublogarithmic space constraints [409]. Hence, in this particular problem, it is possible to overwrite and reuse the memory.

Perhaps it is exactly the necessity to engage—and reuse—one's low-level memory that turns palindromes into such popular and addictive brainteasers.

3.5 Parsing, continued: do brackets matter?

Understanding the role of interiorization and reproduction is crucial for any serious discussion of what is actually happening in teaching and learning mathematics, and it is very worrisome that this cognitive core is so frequently absent from the professional discourse on mathematical education. This is especially true for the discussion of the merits of computer-assisted learning of mathematics, where the use of technology has changed the cognitive content of standard elementary routines which for centuries served as building blocks for learning mathematics.

Typing a command is like saying a sentence, while clicking a mouse is equivalent to pointing a finger in conversation.

And here is a small case study. For some years I had been teaching courses in mathematical logic based on two well-known software packages: SYMLOG [305] and TARSKI'S WORLD [293] (reviews: [107, 115, 118]). SYMLOG used a DOS command line interface which was extremely weak even by the standards of its time, while TARSKI'S WORLD very successfully exploited the graphical user interfaces of Apple and Windows for the visualization of one of the key concepts of logic, a model for a set of formulae (see [8] for the discussion of the underlying philosophy and [403] for underlying mathematics—it is highly non-trivial). Also, TARSKI'S WORLD made a very clever use of games to explain another key concept, the validity of a formula in an interpretation (although the range of interpretations was limited [118]). However, when it came to a written test, students taught with SYMLOG made virtually no errors in the composition of logical formulae, while those taught with TARSKI'S WORLD very obviously struggled with this basic task. The reason was easy to find: SYMLOG's very unforgiving interface required retyping the whole formula if its syntax had not been recognized, while TARSKI'S WORLD's user-friendly formula editor automatically inserted matching brackets. Although TARSKI'S WORLD's students had no difficulty with rather tricky logic problems when they used a computer, their inability to handle formulae without a computer was alarming. Indeed, in mathematics, the ability to reproduce your mental work has to be media-independent. Relieving the students of a repetitive and seemingly mindless task led them to lose a chance to develop an essential skill.

It is appropriate to mention that, in parallel with visualization, there is another mode of interiorization, namely *verbalization*. Indeed, we understand and handle much better those processes and actions which we can describe in words. In naive terms, *typing a command is like saying a sentence, while clicking a mouse is equivalent to pointing a finger in conversation*. The reader would no doubt agree that, when teaching mathematics, we have to incite our students to speak. The tasks of opening and closing matching pairs of brackets, however dull and mundane they may be, activate deeply rooted neural mechanisms for the generation of parsing rules and are crucial for the interiorization of symbolic mathematical techniques.

Sadly, it appears to be acceptable to promote educational software without spelling out what students will lose as a result of its use. In pharmaceutical research, a similar practice would constitute a criminal offence.

I understand that my claims will inevitably provoke the stock response from the promoters of computer-assisted learning: computers are a valuable tool and they help students to save time wasted on routine calculations, allow them to concentrate on deeper conceptual understanding of mathematics, etc. I agree with all that. But I am concerned that the discourse on computer-assisted learning is anti-scientifically skewed and suffers from a cavalier approach to the assessment of the implications for the learner. In medical sciences, promotion of a new medicine without a careful study of its side effects is an academic, regulatory (and, frequently, criminal) offence. In educational circles, it appears to be acceptable to promote a new piece of software for learning a particular chapter of mathematics without spelling out what students will *lose* as a result of its use. Educational software has to be judged on the balance of gains and losses.

3.6 The mathematics of bracketing and Catalan numbers

> We have not begun to understand the relationship between combinatorics and conceptual mathematics.
>
> Jean Dieudonné [24]

The parsing examples we have considered so far have been of a special kind, *binary parenthesizing*; I do not want to venture into anything more sophisticated because even placing parentheses in an expression made by repeated use of a binary operation, such as

$$a + b + c + d,$$

is already an immensely rich mathematical procedure. In various disguises, it appears throughout all of mathematics. There is no better example than Richard Stanley's famous collection of 66 problems on Catalan numbers [406, Exercise 6.19, pp. 219–229] (solutions can be found in [407]). I mention a couple of examples.

The number of different ways to completely parenthesize the formal sum

$$a_0 + a_1 + \cdots + a_{n-1} + a_n \quad (n + 1 \text{ numbers})$$

is called the *n-th Catalan number* and is denoted C_n; it can be shown that

$$C_n = \frac{1}{n+1} \binom{2n}{n}.$$

For example, when $n = 3$, we have 5 ways to place the brackets in $a + b + c + d$, namely:

$$a + (b + (c + d)), \quad a + ((b + c) + d), \quad (a + (b + c)) + d,$$

$$(a + b) + (c + d), \quad ((a + b) + c) + d$$

(following the usual convention, I skip the outermost pair of brackets).

Remarkably, when you count ways to triangulate a convex $(n + 2)$-gon by $n - 1$ diagonals without crossing, you come to exactly the same result:

This mysterious coincidence is resolved as soon as we treat drawing diagonals as taking the sums of vectors

$$\vec{a} + \vec{b} + \vec{c} + \vec{d}$$

going along the $n+1$ sides of the $(n+2)$-gon, with the last side (the base of the polygon) representing the sum:

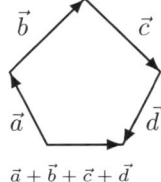

$$\vec{a} + \vec{b} + \vec{c} + \vec{d}$$

Now the one-to-one correspondence between parenthesizing the vector sum and drawing the diagonals becomes self-evident:

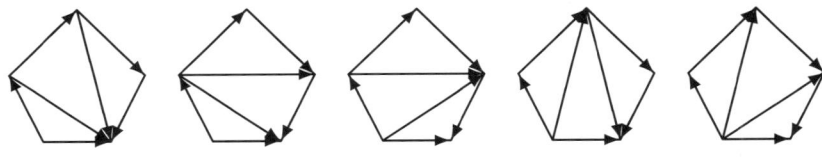

$\vec{a} + (\vec{b} + (\vec{c} + \vec{d}))$ $\vec{a} + ((\vec{b} + \vec{c}) + \vec{d})$ $(\vec{a} + (\vec{b} + \vec{c})) + \vec{d}$ $(\vec{a} + \vec{b}) + (\vec{c} + \vec{d})$ $((\vec{a} + \vec{b}) + \vec{c}) + \vec{d}$

I do not remember the exact formulation of the problem which led me, as a schoolboy, to the discovery of this correspondence between parenthesizing and triangulations, but I remember my feeling of elation—it was awesome.

As a teaser to the reader I give another class of combinatorial objects which are also counted by Catalan numbers. Take graph paper with a square grid, and assume that the unit (smallest) squares have length 1. A *Dyck path* is a path in the grid with steps $(1,1)$ and $(1,-1)$. I claim that the number of Dyck paths from $(0,0)$ to $(2n,0)$ which never fall below the coordinate x-axis $y = 0$ is, again, the Catalan number C_n. I give here the list of such paths for $n = 3$, arranged in a natural one-to-one correspondence with the patterns of parentheses in $a + b + c + d$:

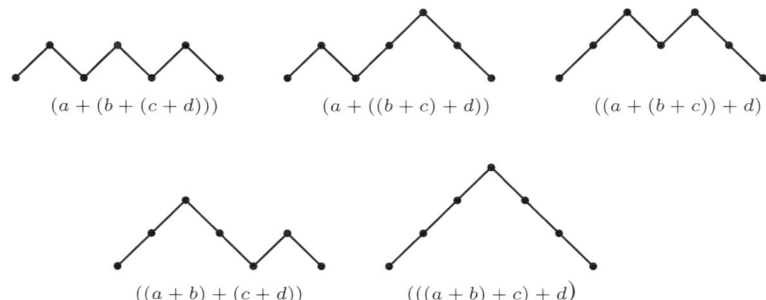

$(a + (b + (c + d)))$ $(a + ((b + c) + d))$ $((a + (b + c)) + d)$

$((a + b) + (c + d))$ $(((a + b) + c) + d)$

Can you describe the rule? Notice that I added, for your convenience, the exterior all-embracing pairs of parentheses; they are usually omitted in algebraic expressions. (Notice also that this correspondence gives, after some massaging, an algorithm for checking the formal correctness of bracketing—so that the algorithm says that the bracketing $(a + (b + c))$ is correct while $(a + b) + c) + ((d + e)$ are not.)

One more example is concerned with n non-intersecting chords joining $2n$ points on the circle:

> *Find a one-to-one correspondence between the 5 chord diagrams and the 5 ways to parenthesize the sum $a + b + c + d$. Hint: there are 3 chords and 3 "+" symbols.*

Again, there are

$$C_n = \frac{1}{n+1} \cdot \binom{2n}{n}$$

different ways to draw the chords. [?]

Richard Stanley makes a wry comment on his list of Catalan number problems [406, pp. 219–229] that, ideally, the best way to solve all 66 problems is to construct directly the one-to-one correspondences between the 66 sets involved, $66 \cdot 65 = 4,290$ bijections in all! It is likely, however, that all 66 sets could be shown to be bijective to one specific set; the set of all rooted trivalent trees with n internal nodes is the most likely candidate for the special role since all 66 sets have a very distinctive hierarchical structure.

This is still not the end of the story: the striking influence of a seemingly mundane structure, grammatically correct parenthesizing, can be traced all the way back to the most sophisticated and advanced areas of modern mathematics research. A brief glance at one of Stasheff's associahedra (Figure 3.2) suggests that they live in the immediate vicinity of Coxeter Theory.[7] Actually, generalized associahedra can be defined for any finite Coxeter group (Stasheff's associahedra being associated, of course, with the symmetric group

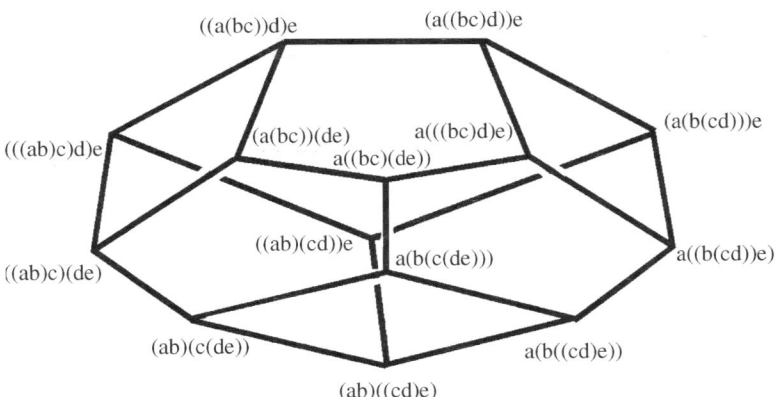

Fig. 3.2. Stasheff's associahedron: the binary parenthesizings of n symbols can be arranged as vertices of a convex $(n-2)$-gon, with two vertices connected by an edge if the corresponding parenthesizings differ by the position of just one pair of brackets.

Sym_n viewed as the Coxeter group of type A_{n-1}); for some recent results see, for example, Fomin and Zelevinsky [351].

3.7 The mystery of Hipparchus

It appears that the importance of parsing has been appreciated by mathematicians and philosophers since ancient times. The following fragment from Plutarch, a famous Greek biographer of the 2nd century A.D., remained a mystery for centuries:

> Chrysippus says that the number of compound propositions that can be made from only ten simple propositions exceeds a million. (Hipparchus, to be sure, refuted this by showing that on the affirmative side there are 103,049 compound statements, and on the negative side 310,952.)

Here Plutarch refers to two prominent thinkers of Classical Greece: the philosopher Chrysippus (c. 280 B.C.–207 B.C.) and the astronomer Hipparchus (c. 190 B.C.–after 127 B.C.). Only in 1994 did David Hough notice that 103,049 is the number of arbitrary (non-binary) parenthesizings of 10 symbols, that is, the number of all possible expressions like

Andrei Zelevinsky,
aged 16

$$(xxxx)((xx)(xx)xx).$$

This numerical observation suggests that, for Chrysippus and Plutarch, "compound" propositions were built from "simple" propositions simply by bracketing.

The mathematics and history of Hipparchus' number is discussed in detail in a paper by Richard Stanley [289]. The number of parenthesizings of n symbols is known as the *Schröder number* $s(n)$; the first 11 values of the Schröder numbers are

$$1, 1, 3, 11, 45, 197, 903, 4279, 20793, 103049, 518859.$$

In 1998, Laurent Habsieger, Maxim Kazarian, and Sergei Lando [269] suggested a very plausible explanation of the second Hipparchus number, of compound statements on the "negative side". They observe that

$$\frac{s(10) + s(11)}{2} = 310,954$$

and, assuming a slight arithmetic or copying error in Plutarch's text, suggest we interpret the compound statements on the "negative side" as parenthesizings of expressions

$$\text{NOT}\, x_1 x_2 \cdots x_{10}$$

under the following convention: the negation NOT is applied to all the simple propositions included in the first pair of brackets that includes NOT. This means that the parenthesizings

$$[\text{NOT}\, [P_1] \cdots [P_k]]$$

and

$$[\text{NOT}\, [[P_1] \cdots [P_k]]]$$

give the same result, and most of the negative compound propositions can be obtained in two different ways. The only case which is obtained in a unique way is when one only takes the negation of x_1. Therefore twice the number of negative compound propositions equals the total number of parenthesizings on a string of 11 elements

$$\text{NOT}\, x_1 x_2 \cdots x_{10}$$

plus the total number of parenthesizings on a string of 10 elements

$$(\text{NOT}\, x_1) x_2 \cdots x_{10}.$$

This, indeed, provides the value $(s(10) + s(11))/2 = 310,954$.

Nowadays, the thinkers of Classic Antiquity do not enjoy the same authority and revered status that they had up to the 19th century. Armed with the machinery of enumerative combinatorics, we may look condescendingly at the fantastic technical achievement of Hipparchus (which became possible perhaps only because he was an astronomer and could handle sophisticated arithmetic calculations, possibly using Babylonian base-60 arithmetic). But I find it highly significant that ancient Greek philosophers, in their quest for understanding of the logical structure of human thought, identified the problem of parsing and attempted to treat it mathematically.

Notes

[1]ONE MORE TEASER. Punctuate:

> Where a previous sentence had had had had had had had had had had had had this sentence contains more.

Continue this inductively to give an arbitrarily large number of perfectly grammatically correct consecutive "hads" in the sentence. (Offered by a commentator on my blog who called himself Ben.)

[2]PARSING RULES. David Pierce has drawn my attention to an interesting question related to parsing rules for mathematical formulae. To what extent is the "infix" notation for binary operations and relations, when the symbol for operation or relation is placed between the symbols for objects, like $a + b$ and $a < b$, made more natural for humans by the nature of their innate grammar generating rules? Or is the predominance of infix notation more of a cultural phenomenon, a fossilized tradition? Why does "reverse Polish notation" (or "suffix" notation) puzzle most people when they first encounter it? In reverse Polish notation, the expression

$$(a + b) \times (c + d)$$

is written as

$$ab + cd + \times.$$

It has serious advantages in computing: when using a hand-held calculator designed and programmed for the use of reverse Polish notation, one is not troubled with saving the intermediate results into the memory; this is done automatically. On an ordinary calculator, one has to save the intermediate result $a + b$ when calculating $(a + b) \times (c + d)$. Notice that infix notation does not generalize from binary to ternary operations, and ternary operations and relations are not frequently found in mathematics. Is that because our writing is linear, reflecting the linear nature of speech? Words denoting ternary or higher arity relations are infrequent in human languages. The predicate "a is between c and d" is a noticeable exception in English. Interestingly, the "betweenness" relation among points on a line was famously absent from Euclid's axiomatization of geometry (see Section 11.4).

[3]SINGULAR, DUAL, TRIAL, PAUCAL. . . . In general, languages tend to treat numbers from 1 to 4 differently; see [196]. Owl remarked that traces of the dual category still can be found in Russian (and it is still present in

Slovenian. Apparently, dual was mostly purged from Russian in the language reform of Peter the Great.

Barbara Sarnecka wrote to me:

... usually the options are

(a) Singular (1)/Plural (2+),

(b) Singular (1)/Dual (2)/Plural (3+),

(c) Singular (1)/Dual (2)/Trial (3)/Plural (4+),

(d) Singular (1)/Dual (2)/Paucal (approximately 3–4)/Plural (approximately 5+).

Anyway, Russian is the only language I know of where the dual and paucal categories have been merged into one, so that is quite interesting. Is it possible that there was, earlier, a singular/dual/paucal/plural system, and that Peter [the Great] tried to simplify it by combining the dual and paucal categories?

I would be happy to learn more about pluralities—although this theme leads well beyond the scope of my book. In particular, I was intrigued by a comment from my Hungarian colleague that, in the Hungarian language, a plurality marker is present in nouns when no specific numeral is used with the noun.

As I have already mentioned in the main text, I shall return to the discussion of thresholds for pluralities in Section 4.1.

[4]CANON IN POETRY. My dear old friend Owl reminded me that canon can be found in poetry, where it is sometimes used with a totally mesmerizing effect:

В посаде, куда ни одна нога
Не ступала, лишь ворожеи да вьюги
Ступала нога, в бесноватой округе,
Где и то как убитые спят снега, –

Постой, **в посаде куда ни одна**
Нога не ступала, лишь ворожеи
Да вьюги ступала нога, до окна
Дохлестнулся обрывок шальной шлеи.
(Boris Pasternak)

[5] PALINDROMES AND COXETER GROUPS. My "palindrome" formulation of the Coxeter-Tits Theorem is one of many manifestations of a *cryptomorphism*, the remarkable capacity of mathematical concepts and facts for translation from one mathematical language to another; see more on that in Section 4.2. I recall again that, in this book, I have adopted a "local", "microscopic" viewpoint. Although the "palindrome theory" is of little "global" value for mathematics, it demonstrates some interesting "local" features of mathematics.

[6]Without the assumption about the finiteness of the number of palindromes, the Palindrome Representation Theorem is still true if we accept mirrors in non-Euclidean spaces. Section 5.1 contains some examples of mirror systems in the hyperbolic plane.

[7]An elementary construction of associahedra can be found in Loday [381].

4

Simple Things

I remember simple things.
I remember how I could not understand simple things.
This makes me a teacher.
Hovik Khudaverdyan

4.1 Parables and fables

And he spake this parable unto them, saying,
What man of you, having an hundred sheep,
if he lose one of them,
doth not leave the ninety and nine in the wilderness . . .
Luke 15:3–4

Philosophers of mathematics find it useful to look at mathematical texts as narratives (see, for example, David Corfield [17] and Robert Thomas [92]). Indeed, even an average, run-of-the-mill mathematical paper has a multi-layered structure of complexity comparable with that of a serious novel, like *War and Peace* or *Ulysses*. The analogy is, however, much deeper; for me, its most appealing aspect is a parallelism between the development of a character in a novel or play and specialization of an abstract mathematical structure.

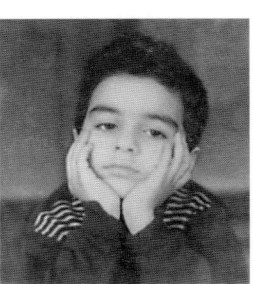

Hovik Khudaverdyan, aged 6

However, I wish to follow the principal line of this book and to look at something smaller and simpler than a novel.

So-called *mathematical folklore* is virtually unknown outside professional circles: it is the corpus of small problems, examples, brainteasers, jokes, etc., not properly documented and existing mostly in oral tradition. It is a small universe on its own; but in all its diversity, one can easily notice the prominent role of fables or parables, that is, general statements (or problems) which are intentionally set in the least general terms or illustrated by a simple, highly specialized example.

In the written tradition, there is at least one famous parable, the celebrated *Pigeonhole Principle*:

> If you were to put 6 pigeons in 5 holes, then at least one hole would contain more than one pigeon.

In the Russian mathematical literature, the Pigeonhole Principle is known by the name of the *Dirichlet Principle*.[1] The name emphasizes its pedigree and status; however, the principle itself is usually formulated in terms of 6 rabbits and 5 hutches.

Robert Thomas,
aged 13.
© Estate of Mary G.
Thomas

What always struck me since the time when I first encountered the Pigeonhole Principle, was the persistence of the numbers 6 and 5 in its formulation. I felt that the choice was somehow very precise and convincing; I have sometimes seen alternative formulations, as a rule, with similar small numbers, but they somehow looked less attractive.

I can now see a possible explanation of the persistence of the "6/5" formulation in works on the neurophysiology of counting. Indeed, it is now an established fact that the mechanisms of perception of small ensembles of objects are very different from counting larger ensembles. Up to 5 objects, we have *subitizing*, i.e., "suddenizing", immediate perception of the quantity which does not interfere with our ability to keep track of each individual object [171]; subitizing starts to fail at 6 objects, and, as a rule, has to be switched to counting when we have 7 or more objects.[2]

The cultural significance of the subitizing/counting threshold was understood early on. Comments from one of the first experimental psychologists who wrote about subitizing and a related phenomenon, short-term memory (George A. Miller [206], 1956) were interesting for being both suggestive and very cautious:

> And finally, what about the magical number seven? What about the seven wonders of the world, the seven seas, the seven deadly sins, the seven daughters of Atlas in the Pleiades, the seven ages of man, the seven levels of hell, the seven primary colors, the seven notes of the musical scale,[3] and the seven days of the week? What about the seven-point rating scale, the seven categories for absolute judgment, the seven objects in the span of attention, and the seven digits in the span of immediate memory? For the present I propose to withhold judgment. Perhaps there is something deep and profound behind all these sevens, something just calling out for us to discover it. But I suspect that it is only a pernicious, Pythagorean coincidence. [4]

Unlike subitizing, in counting, our attention moves from one object to another. On the other hand, and somewhat surprisingly, experimental studies (using PET, positron emission tomography scans) have failed to find differences in the neurophysiological activities of the brain in subitizing and counting [216].

Even if it turns out that subitizing and counting are both implemented by the same system of neuron circuits, they should correspond to two different modes of its activity, with some kind of a phase transition between the two. The task of mentally putting 6 pigeons (recall, the borderline value) into 5 holes should, therefore, put the system in the critical zone. It is tempting to suggest that the criticality of the 6/5 combination may provoke the strongest response, leading to a new pattern of interaction (or interference) between neuron circuits.

David Pierce,
aged 13

Again, this is my speculative guess based on introspection, but I believe that mathematicians may like the classical "6/5" formulation of the Pigeonhole Principle because they hear, deep within themselves, a subtle click made by a mathematical concept attaching itself to the neuron circuitry of their brains.

After I wrote this paragraph, I come across the following excerpt from Vandervert [242, p. 87]:

> The experience of intuition "is" the feel of the entrainment, so to speak, of the neuro-algorithms of perception with the newer ontogenetic neural subcircuitry retoolments (Edelman [[175]]) that undergird mathematical discovery. We might speculate that the "aha" experience and exclamation occur upon *recognition* of the locking-in of the entrainment of the two systems of algorithms.

I am in agreement with this position; however, I prefer to express the same ideas is simpler words, leaving it to experimental neuroscientists to develop an appropriate terminology. Also, I would rather avoid the use of the word "intuition" as both excessively general and, at the same time, restricted to the process of mathematical discovery. The "locking-in" can be much more frequently found in routine everyday activities such as understanding and digesting other people's mathematics. It can definitely be found in the act of accepting a proof. Remember Coxeter's proof of Euler's Theorem (Section 2.3); do you hear that click in your brain?

It is likely that some mental objects have a higher degree of affinity to the hardwired structures of human cognition and anchor themselves more easily than others, while more sophisticated ones require the mediation of mental objects which have already been interiorized. For the moment, let us treat this as no more than a metaphor for the inner working of a mathematician's brain.

As I discuss in more detail in Section 12.5, computer science and complexity theory might provide some hints for the further development of this metaphor, for predicting or explaining why certain objects are easier to interiorize than others. It is worth mentioning that computer scientists and cognitive scientists have already started to think about abstract models of counting and subitizing. For example, a possible model of counting is discussed by da Rocha and Massad [222], who claim that such models can be constructed from the so-called Distributed Intelligent Processing Systems. Peterson and Simon [215] claim to have an executable model of subitizing of up to 4 objects (apparently, it is available for download from the Internet). But what I would really like to see is an abstract model of counting (possibly, a further development of [222] or [215]) which explains the subitizing/counting threshold (and the "6/5" formulation of the Pigeonhole Principle) at the "software level", thus accounting for the indiscernibility of these two activities at the physiological, "hardware" level.

In short-term memory, we also have a threshold of 7 ± 2 similar to that of the subitizing threshold: people usually can memorize a 7-digit telephone number, but they encounter serious difficulty with 10-digit numbers. Mathematical analogues of memory are easier to formulate than that of subitizing. Since memory is an adaptive, ever-changing, and dynamic system, stable patterns in *dynamical systems* (the words "dynamical system" are now understood as a precise mathematical term) appear to be natural candidates for mathematical phenomena whose behavior might be analogous to the behavior of human memory. Can the 7 ± 2 threshold be found in mathematical dynamical systems?

I quote, at length, from a paper by Paul Glendinning [185] who explains recent works by Kaneko aimed at exactly this elusive target: find a natural 7 ± 2 threshold in the behavior of dynamical systems [197, 198].

> Kaneko's starting point is the idea of an attractor of a dynamical system. Classically attractors are thought of as invariant sets which 'attract' nearby points. That is, there exists an open neighbourhood of the set such that any solution with initial conditions in this neighbourhood eventually tends to the invariant set. There are all sorts of variants on this definition, but the defining feature of the attractor is a neighbourhood on which some property of attraction holds. Twenty years ago Milnor [[386]] pointed out that this is a topological definition and introduced a measure-theoretic definition in which the open neighbourhood is replaced by a set of positive measure locally (or, again, a variant of this idea). The difference between the definitions is one of how to give the words 'lots' or 'most' mathematical meaning—either in a topological sense (open neighbourhoods) or in a measure-theoretic sense (positive measure). The term Milnor attractor is now used to describe an

attractor which attracts a large set in measure but not in topology. The important point here is that in any neighbourhood of a Milnor attractor, there are points which move away from the attractor, and which may be in the basin of another attractor. This gives Milnor attractors an interesting property: for a topological attractor *all* solutions close enough to the attractor are attracted, which is a sort of stability, whilst for a Milnor attractor there are points arbitrarily close to the attractor which move away from the attractor.

Suppose now that a 'memory' is an attractor of a dynamical system (the brain: neurons, etc.). To be stable to perturbations, i.e. to be a useful memory, it is natural to ask that a memory should be a topological attractor rather than a Milnor attractor. The memory of a telephone number of N digits may be represented by systems in $\mathbb{R}^N$ (although more subtle questions about information content could be explored). Kaneko [[198]] considers a 'prototype' system of globally coupled maps and shows that the proportion of points which tend to Milnor attractors and hence the proportion which correspond to poor memory increases until $N \sim 7$ and then plateaus. In other words, in order to minimize the proportion of easily forgotten states one should keep the dimension below 7. He has a rough, but appealing, argument to support this for the systems he considers which basically comes down to a combinatorial balance between $(N-1)!$ and 2^N, the balance being at $N \sim 5$.

Shuji Ishihara and Kunihiko Kaneko [[197]] extend this idea to a more conventional neural net model. This has N inputs $x_i(0)$, $i = 1, \ldots, N$, and a feed forward mechanism which passes information through successive layers indexed by ℓ where

$$x_i(\ell + 1) = \tanh\left(\frac{\beta}{\sqrt{N}} \sum_k a_{ik}(\ell) x_k(\ell)\right)$$

and where $a_{ik}(\ell)$ are chosen randomly from a Gaussian distribution with standard deviation 1. The tanh function acts as a sigmoidal on-off switch for large enough β, with solutions approaching values close to ± 1. If $0 < \beta < 1$, then solutions decay to zero. There is an intermediate range of β where the behavior of the system depends significantly on N. If N is less than about (you've guessed it) 7, then the output at layer L assumes only a small number of distinct values and there is a clear separability of inputs (Ishihara and Kaneko work with $L = 30$, but the principle is independent of the depth of layers used). If N is larger, then the dynamics as a function of layer is chaotic, and small changes in the input (which we would hope should stabilize) create large differences in the output. This critical changeover in behaviour is striking, and once again suggests that there is a critical size of systems above which information becomes garbled.

The subitizing and short memory thresholds are just two of many problems which make me yearn to see the dawn of *cognitive metamathematics* which would turn the "software/hardware" metaphor into a theory. I believe that these are not unrealistic expectations. The development of brain scan techniques appears to have reached a level where at least some of the ideas mentioned in this book can, with due effort, be made into experimentally refutable conjectures.

However, we shall still have only isolated experiments and observations until mathematicians start, in earnest, to develop mathematical models of mathematical cognition—this is where the true cognitive metamathematics will be born.

It is time now to return to the discussion of the 4/5 threshold in use of plurality markers in Russian; see Section 3.2. I mention there an alternative explanation of their appearance: the predecessor of the Indo-European language had numerals formed from base 4 (fingers of the hand) with thumb marking the next register [438].

Numerological theories (like conspiracy theories), however, are famous for their resilience. Indeed, there still remains a possibility that phase transitions in behavior of attractors in dynamical systems would provide a *uniform* explanation for *both* the subitizing threshhold 5/6 and the fact that we have 4+1 fingers. This phase transition can manifest itself in a variety of ways:

(a) in differentiation of cells in embryogenesis, where a $4 + 1$ finger anatomy can happen to be the easiest to achieve,
(b) in a relative ease of the neural control of complex movements of a hand with $4+1$ fingers in comparison with other designs (here, nature tends to prefer simple solutions—this has been already observed by neuroscientists [153]),
(c) in the dynamics of patterns of activation of neural paths in image processing in humans which imposes the subitizing/counting threshold,
(d) finally, in the architecture of short-term memory and its influence on word processing in humans.

It is possible, of course, that some other mathematical theory can work in place of the theory of dynamical systems. So far, highly speculative applications of dynamical systems bring to mind an old adage:

if all you have is a hammer, everything looks like a nail.

4.2 Cryptomorphism

The reader will find in this book a number of reformulations of well-known theorems and theories intentionally made in a "toy" language—a typical example is my "palindrome" formulation of the Coxeter–Tits Theorem in Section 3.4; see endnote 5 in Chapter 3. These are instances of *cryptomorphisms*, the remarkable capacity of mathematical concepts and facts for a faithful translation from one mathematical language to another. Such translations constitute an important but underrated part of mathematical practice; they remain virtually unknown outside professional circles. It can be surmised that many teachers of mathematics view "multiple representations" of mathematical objects as a hindrance; see a more detailed discussion in Section 7.5.

One of the reasons why the "language" aspect of mathematics is ignored in mainstream mathematical education is that most translations, as with much of mathematical work generally, never make it from scratch paper to publication. The situation is different in "Olympiad" or "competition" mathematics which pays more attention to what is happening on scratch paper and which also needs a steady flow of new original (or attractively disguised old) problems for higher level competitions.

To give you the flavor, here is a "double" problem from a classical Olympiad problem book [265, Problem 145]. [?]

Solve both versions of the problem.

a. Two people play on a chessboard, moving, in turn, the same piece, the King. The following moves are allowed: one square left, one square down, or the diagonal move left-down. The player who places the King in the leftmost square in the first (bottom) row wins. At what initial positions does the first player have a winning strategy?[5]

b. Two players take, in turn, stones from two heaps. They are allowed either to take one stone from one heap or to take one stone from both heaps. The player who picks the last stone wins. At what initial number of stones in the two heaps does the first player have a winning strategy?

Is all that just a game? Does the bewildering variety of mathematical languages which can express the same fact matter? Let us listen to two expert opinions.

4.2.1 Israel Gelfand on languages and translation

My position on the issues of cryptomorphism and "multiple presentation" is greatly influenced by my conversations with Israel Gelfand. He once said to me:

> Many people think that I am slow, almost stupid. Yes, it takes time for me to understand what people are saying to me. To understand a mathematical fact, you have to translate it into a mathematical language which you know. Most

mathematicians use three, four languages. But I am an old man and know too many languages. When you tell me something from combinatorics, I have to translate what you say in the languages of representation theory, integral geometry, hypergeometric functions, cohomology, and so on, in too many languages. This takes time.

Mathematical languages unstoppably develop towards an ever increasing degree of compression of information.

It is amusing to watch how fellow mathematicians, not accustomed to the peculiarities of Gelfand's style, speak to him the first time. Very soon they become bewildered at why he insists on their giving him really basic, everyone-always-knew-it kinds of definitions; then they are taken aback when he becomes furious at the merest suggestion that the definition is easier to write down than to say orally ("I know, you want to cheat me; do not try to cheat me!"). The next morning, their second conversation is usually even more entertaining, because Gelfand starts it with the demand to repeat all the definitions; then he proceeds by questioning everything which was agreed upon yesterday, and eventually settles for a definition given in a completely different language.

I have observed such scenes many times and came to the conclusion that, for him, a definition of some simple basic concept, or a clear formulation of a very simple example, is a kind of synchronization marker which aligns together many different languages and makes possible the translation of much more complex mathematics.

4.2.2 Isadore Singer on the compression of language

Mathematicians are so sensitive to mathematical language issues because they can see dramatic changes in the languages used over their working lifetime.

Another aspect of the Babel of mathematical languages, their unstoppable development towards an ever increasing degree of compression of information, is succinctly expressed by Isadore Singer in a recent interview [77]:

I find it disconcerting speaking to my young colleagues, because they have absorbed, reorganized, and simplified a great deal of known material into a new language, much of which I don't understand. Often I'll finally say, "Oh; is that all you meant?" Their new conceptual framework allows them to encompass succinctly considerably more than I can express with mine. Though impressed with the progress, I must confess impatience because it takes me so long to understand what is really being said. [77, p. 231]

One of the reasons why research mathematicians are so sensitive to mathematical language issues is that they can see dramatic changes in the languages used over their working lifetime.

4.2.3 Cognitive nature of cryptomorphism

Returning to the running example of this book, we see that palindromes and mirrors are, essentially, cryptomorphic objects. They are sufficiently basic and "atomic" to belong simultaneously to two different realms of cognition, the verbal/symbolic and the visual, although the status of palindromes is clearly borderline: to appreciate a palindrome, you have to *see* the symmetry of its presentation in type. Remember the peculiar typesetting of "ABBA" on their posters? And a question to experimental psychologists: do blind people (especially if they are blind from birth) aesthetically appreciate palindromes when they read them in Braille? In more general terms, is there any significant difference in the perception of symmetry by blind people from that of the sighted?

A similar observation also appears to be valid in the case of *musical palindromes*, where the visual symmetry of the score is of importance; see the analysis of symmetry in music in Wilfrid Hodges' paper *The geometry of music* [51]. (In particular, Hodges discusses the paradoxical results of playing a recording of palindromic music (Haydn's *Menuet al Rovescio* for piano) *backwards*: the individual notes, as produced by musical instruments, are not reversible in time. For example, piano notes start with a bang and then fade away.

I believe that no cognitive theory of mathematical practice can be complete without a discussion of the nature and role of the cryptomorphism. The crucial point here is the understanding of relations between the conscious use of the cryptomorphism and the neurophysiological interaction between various structures of our brain, for example, between the visual and verbal modules.

Even isolated glimpses into what is happening in our brains—for example, case studies of people who compensate for the lack of function in certain sensory faculties—are of great interest (like the case study of the compensation for saccadian eye movements [182] briefly mentioned earlier; see Section 2.2).

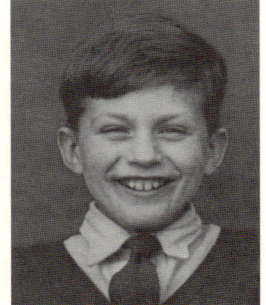

Wilfrid Hodges, aged 10

> No cognitive theory of mathematical practice can be complete without a discussion of the nature and role of cryptomorphisms.

I would hope that, eventually, it will become possible to identify certain elementary mathematical structures and concepts which are sufficiently atomic and are especially nimble in moving from one cognitive structure to another; essentially, they can be viewed as elements of the human cognitive system. Let us call these elementary particles of mathematics *mathlets*. I want to suggest, not so much as a conjecture but rather as the starting point of a discussion, that the notion of cryptomorphism, as part of mathematical thinking, is supported, at the level of basic cognitive structures, by the transcription of mathlets from one cognitive subsystem of our mind to another. I discuss some examples in the next section.

4.3 Some mathlets: order, numerals, symmetry

Everything should be made
as simple as possible, but not simpler.
Albert Einstein

4.3.1 Order and numerals

Order is almost definitely one of these mathlets, elementary particles of mathematics. Frank Smith [132] points out that the understanding of order precedes counting and the understanding of numerals. What struck me most in Frank Smith's lively discussion of pedagogical practice is that children need, as the foundation of all their arithmetical (and more generally, mathematical) activities one, just one, *fixed linearly ordered discrete set explicitly produced by a simple recursive rule.*

His thesis is supported by Susan Carey [162, 67]:

> Research suggests that it is not until *after* children have worked out how the count list represents number [...] that they know which analog magnitudes correspond to which numbers above five in their count list.

And also:

> Children may here make a wild analogy—that between the order of a particular quantity within an ordered list, and that between this quantity's order in a series of sets related by additional individuals. These are two quite different bases of ordering—but if the child recognizes this analogy, she is in the position to make the crucial induction:

For any word on the list whose quantificational meaning is known, the next word on the list refers to a set with another individual added. Since the quantifier for single individuals is 'one,' this is the equivalent to the following induction: If number word X refers to a set with cardinal value n, the next number word in the list refers to a set with cardinal value $n + 1$.

(See Sections 5.2 and 5.3 for the discussion of how infinity creeps into this inductive process.)

One may argue that the analogy referred to by Susan Carey is, more precisely, between the *position* of a *word* for a quantity, in a list of such words, and the *size* of a set whose quantity is named by that word—this should be more likely at least at the early stages of learning numbers and agrees better with the phenomenon of "one-, two-, four-knowers"; see Section 3.2.

All other ordered sets which appear, explicitly or implicitly, in children's mathematical practice (and just everyday life) are understood by reference to that distinguished ordered set of (verbal) numerals, or it can be immediately reduced to it, even if they appear in visual perception (objects on the line), tactile and spatial perception (leftmost-left-right-rightmost), or as the pitch and relative height of tones in music (although in pitch perception, the order frequently happens to be *cyclic*; many people perceive tones which differ by octave as identical), or as the relative measure of muscular tension (weighing objects in the hand).

This list of manifestations of order in the human senses can be easily continued and can be related to a huge body of psychological literature.

There are reasons to believe that the correspondence between verbal numerals and the intuitive perception of order is not inborn, but learned. Children have to bridge gaps between "approximate" arithmetic, estimation activities (even after they are verbally expressed) and "exact" symbolic arithmetic. The term "approximate" arithmetic refers to the ability to distinguish between quantities—by choosing, say, a box with a larger number of bananas; such ability can be found in very young children as well as in apes and other animal species. This is how Dehaene et al. describe it:

> [...] even within the small domain of elementary arithmetic, multiple mental representations are used for different tasks. Exact arithmetic puts emphasis on language-specific representations and relies on a left inferior frontal circuit also used for generating associations between words. Symbolic arithmetic is a cultural invention specific to humans, and its development depended on the progressive improvement of number notation systems. [...]

Approximate arithmetic, in contrast, shows no dependence
on language and relies primarily on a quantity representation implemented in visuo-spatial networks of the left and
right parietal lobes. [170, p. 973]

The transfer of order between the different sensor mechanisms
can be conscious or subconscious. One of more striking examples
of subconscious transfer is given by Leontiev's experiments [202,
pp. 193–218] on training tone-deaf adults to distinguish the pitches
of sounds by developing a correlation between pitch and muscular
tension. Interestingly, visual perception played the role of interme-
diary: the subjects had to pull a lever and equalize two gauges,
one for pitch and another for the force applied. After some training of that kind, subjects attained a reasonable ability to recognize
pitch even without gauges and levers. Their test performance de-
teriorated, however, when the devious experimenter engaged the
subjects' hands (or feet, if, in the training period, the lever was
pushed by foot) in some activity. This example is even more puzzling because my applied mathematics colleagues told me that the
sound receptors in our inner ear are actually doing, at the hardware level, a Fourier transform [227]. At what point do the brains
of tonally challenged people start to ignore hardware readings?

In learning arithmetic, the understanding of ordinal numbers
precedes the understanding of cardinal numbers. Might one suggest that the early-childhood distinction between ordinals and cardinals is mirrored in the mathematical difference in the transfinite
realm?[6] When learning set theory, ordinal numbers are more intuitive than cardinals—perhaps because a fragment of the theory of
ordinals concerned with countable ordinals is relatively accessible
and has many elementary aspects. Historically, Georg Cantor introduced ordinal numbers first, and only later came to the concept of
the cardinality of a set. I briefly mention ordinals in Section 6.2.2;
see, in particular, Figure 6.1. One might also note that the cardinal
number one need not correspond to the first ordinal; the first ordinal could be zero, or minus sixteen. In English, the words 'first' and
'second' have no etymological connection with 'one' and 'two'. The
first is the foremost, the thing in front of everything else; the second is the following. In ancient Greek language, numbers started
with two. In modern English, the word 'number' as used in expressions like "we have a number of options" has an implicit meaning
of "more than one".

I return to the discussion of order in Section 6.3.

4.3.2 Ordered/unordered pairs

The fact that order is built into human languages becomes especially transparent when we notice that our language is not good at

describing an *unordered pair* of two objects: we cannot name two objects without giving preference to one of them, simply because one of the names will precede the other in time.

Vladimir Uspenky [455] finds this aspect of human languages essential for understanding their logical structure and laments that it has not attracted the due attention of researchers. I quote at length Uspenky's analysis of the use of an unordered pair as opposed to an ordered pair, in a scene from Dostoevsky's *Idiot*. Uspensky quotes a conversation between Prince Myshkin and Lebedeff in which Prince Myshkin asks Lebedeff's given name and patronymic (the latter is formed from the father's given name; in Russia, it is customary to use the name and the patronymic as a polite form of address).

> "Listen to me, Lebedeff," said the prince [...] "By the way—excuse me—what is your Christian name? I have forgotten it."
> "Ti-Ti-Timofey."
> "And?"
> "Lukianovitch."
> Everyone in the room began to laugh.
> "He is telling lies!" cried the nephew. "Even now he cannot speak the truth. He is not called Timofey Lukianovitch, prince, but Lukian Timofeyovitch. Now do tell us why you must needs lie about it? Lukian or Timofey, it is all the same to you, and what difference can it make to the prince? He tells lies without the least necessity, simply by force of habit, I assure you."
> "Is that true?" said the prince impatiently.
> "My name really is Lukian Timofeyovitch," acknowledged Lebedeff, lowering his eyes, and putting his hand on his heart.
> "Well, for God's sake, what made you say the other?"
> "To humble myself," murmured Lebedeff.
> "What on earth do you mean?" [...] cried the prince...
> (Translated by Eva Martin)

Humility may show in man's desire to give as little information about himself as possible—because he does not think that this information is of any value to others. Explicitly asked about his name and patronymic, Lebedeff cannot give just a partial answer "Lukian", since it would not be polite. The complete answer is the ordered pair (Lukian, Timofey); obviously, an unordered pair

$$\{\text{Lukian}, \quad \text{Timofey}\}$$

(which is, of course, the same as {Timofey, Lukian}) contains less information than the ordered pair. Poor Lebedeff is trying to convey just an unordered pair of names {Lukian, Timofey}, but, of course, he cannot; his attempt is hampered by the linear nature of human speech which imposes order against the will of the speaker.

Uspensky then quotes a scene from Nikolai Gogol's *Dead Souls*: Chichikov's visit to Manilov.

> ... it is time that we returned to our heroes, who, during the past few minutes, have been standing in front of the drawing-room door, and engaged in urging one another to enter first. ... Finally the pair entered simultaneously and sideways; with the result that they jostled one another not a little in the process. (Translated by D. J. Hogarth)

We see that Gogol avoids calling his characters by names: if he did so, he would create an undesirable expression that one of the friends was more persistent in his politeness than the other.

4.3.3 Processes, sequences, time

Closely related to order is our innate facility for thinking about processes or sequences of actions which take place in time. Indeed, phonetic parsing takes place in time; our writing is mostly linear, one symbol at a time; we read two-dimensional flowcharts by tracing paths, arrow by arrow; reviewers and editors of mathematical papers grudgingly tolerate nested subscripts and superscripts, like x_{i_n}, but branching combinations of nested superscripts and subscripts like

$$x_{i_n}^{m_k^l}$$

are considered to be contrary to the norms of good writing.

Thurston stresses that our intuition of process can be

> ... used to good effect in mathematical reasoning. One way to think of a function is as an action, a process, that takes the domain to the range. This is particularly valuable when composing functions. Another use of this facility is in remembering proofs: people often remember a proof as a process consisting of several steps. In topology, the notion of a homotopy is most often thought of as a process taking time. Mathematically, time is no different from one more spatial dimension, but since humans interact with it in a quite different way, it is psychologically very different. [94, p. 165]

4.3.4 Symmetry

Finally, I return to the running example and one of the principal themes of the book. Our analysis of the human's perception of symmetry in Chapter 2 allows us to conjecture that symmetry, and especially bilateral (mirror) symmetry, is another important elementary mathlet.

The discussion of symmetry, in its various disguises, and its cognitive nature, is spread all over the book.

4.4 The line of sight and convexity

Do androids dream of electric sheep?
Philip K. Dick

Let us return for a minute to our discussion of the electromagnetic imaging in fish which we started in Section 2.1, page 25. Which immediately intuitive mathematical concepts would become less intuitive if, instead of sight, humans used electric sensing of the kind used by Nile elephant fish *Gymnarchus niloticus*? The crucial difference is that we would loose the concept of the *line of sight*, the archetypal straight line of our geometry. Indeed, electric sensing would allow us to "see" *around objects*; closer objects would not obscure the view of more distant ones.

It is worth remembering that Euclid (or a later editor of *Elements*) defines a straight line as

a line that lies evenly with its points.

It makes sense to interpret this definition as meaning that a line is straight if it collapses to a point when we hold one end up to our eye. Therefore, a straight line is a line of sight![7]

Chapter 10 contains a number of simple problems illustrating the relations between the concept of line of sight and that of *convexity*. Recall that a subset X of the n-dimensional Euclidean space $\mathbb{R}^n$ is *convex* if it contains, with any points $x, y \in X$, the segment $[x, y]$ (Figure 4.2).

The class of compact convex bodies can be characterized in terms of the relation

"A (partially) obscures B".

Indeed, the set of compact convex bodies[8] in $\mathbb{R}^n$ can be characterized as the maximal possible collection $\mathcal{C}$ of compact bodies in $\mathbb{R}^n$ such that

- $\mathcal{C}$ is closed under all rigid movements and similarity transformations.
- If you look at two non-intersecting bodies A and B, both taken from the collection $\mathcal{C}$, and A partially obscures B in your field of vision, then B cannot (partially) obscure A (therefore the relation "A obscures B" is *antisymmetric* in the class of convex bodies).

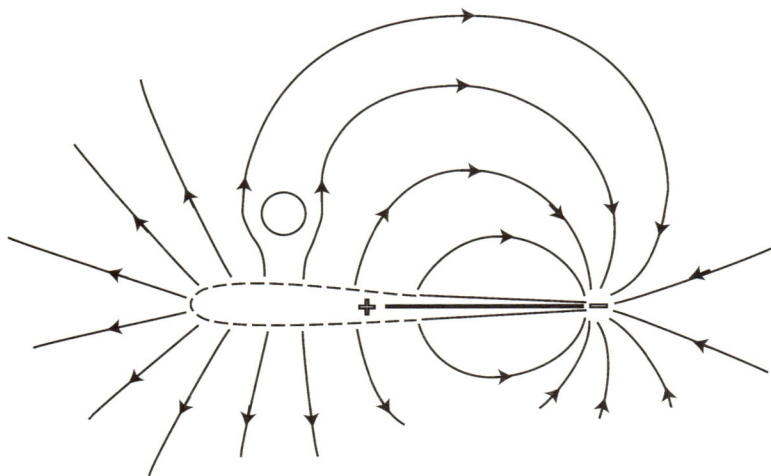

Fig. 4.1. Schematic representation of the principles of active electrolocation in fish. The electric organ (solid black bar) gives rise to a dipolar
field pattern around the fish. The electric current follows the local field
lines (solid lines with arrows). A non-conducting target object (circle)
perturbs the flow of electric current, causing a local decrease in current
density near the object. This decrease in current density translates into a
decrease in the transdermal voltage across the skin near the object. The
spatial pattern of the transdermal voltage across the sensory surface represents the electric image of the object.

The body surface is covered with thousands of sensors that measure
local changes in voltage across the skin (transdermal potential). You can
think of each sense organ as an electrode pair that measures the potential difference across the highly resistive fish skin.

One interesting thing is that the field pattern and receptor distribution allow the fish to detect objects in all directions—an omnidirectional
sensing capability.

Another observation is that the density of sense organs is rather high
(several per square millimeter), suggesting that the system may be good
for spatial localization of small prey targets that are just a few millime-
ters in diameter.

Also, the sensor density tends to be higher near the mouth, suggesting
that this is the region of the body where fine spatial localization might be
most important.

(Redrawn and quoted from Nelson [209], with kind permission of
Springer Science + Business Media; the drawing originates in Heiligenberg [190].)

| Prove it! | I leave the proof to the reader as an exercise. [?]

We have already discussed the prominent role of order among
basic "built-in" mathematical concepts of the human mind. A strict

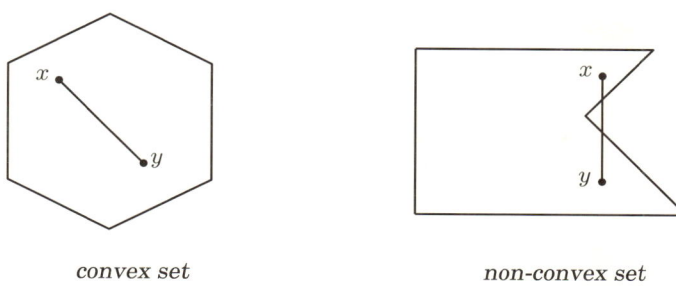

convex set   *non-convex set*

Fig. 4.2. Convex and non-convex sets.

order $<$ is an anti-symmetric relation (that is, statements $x < y$ and $y < x$ cannot hold simultaneously). The other part of the definition of strict order is that it is a transitive relation (that is, $x < y$ and $y < z$ implies $x < z$).

Now we have one more piece of evidence of the special role of order in mathematical cognition. Indeed it appears that the human visual processing system, when dealing with convex objects, frequently assumes that the "A obscures B" relation is not just anti-symmetric but is also transitive (and hence a strict order): if body A is in front of body B and B is in front of C, then A is in front of C. It is a systematic error of our brains, and it can be seen in many visual paradoxes with non-existing objects. For example, have a look at Figure 4.3 and try to decide which parts of the contraption are closer to the viewer and which are more distant.

The chapters of geometry dealing with convex bodies contain a number of results which are surprisingly intuitive and self-evident. Here is one example:

> A convex polytope (that is, a convex and bounded polyhedron) is the convex hull of its vertices (that is, the smallest convex set containing the vertices).

R. T. Rockafellar in his fundamental treatise on convex geometry emphasized the paradoxical status of this statement [398, p. 171]:

> This classical result is an outstanding example of a fact which is completely obvious to geometric intuition, but which wields important algebraic content and is not trivial to prove.

David Henderson, aged 15. Photo supplied by D. Henderson.

In relation to convex bodies and their properties, our intuition about convex bodies is both very powerful and very misleading— perhaps because we are excessively confident in our judgement—

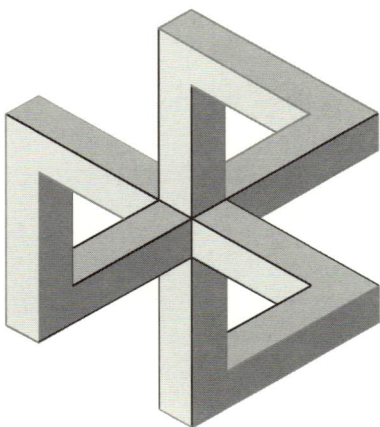

Fig. 4.3. This impossible geometric object is a combination of three Penrose triangles. I showed it to a few fellow mathematicians—all agreed that the picture makes for uncomfortable viewing: one's brain continues to seek *three-dimensional symmetries* even when one knows that the object is not just highly asymmetrical—it is non-existent! As an image in the plane, the object has rotational symmetry of order 3 (if one ignores the difference in shades of color on its facets).

and I refer the reader to Chapter 10 for more examples of both feasts and failures in our intuition.

4.5 Convexity and the sensorimotor intuition

There is one more aspect of the human intuition of convexity: it is immediate to us in our tactile and sensorimotor perception. It could be illustrated by the mirror grinding problem, Section 2.3, page 30. It is immediately clear that when we grind one piece of glass over the other, the surface of one piece becomes concave, while the surface of the other convex. Just perform a small experiment: close your eyes and move you hands as if you are indeed grinding the mirrors. When we change pieces, the concavity and convexity alternate; with three pieces, the concavity and convexity have to cancel each other and the surfaces become both concave and convex, that is, flat. This gives an answer to the first grinding problem, problem (A) on page 30. However, a formal rigorous proof is probably very hard.

Problem (B) is trickier. Since we do not have many names for surfaces in our language, it is easy to assume that the answer is probably a sphere. Yes, indeed; but the reader would probably agree

Fig. 4.4. *The Magpie on the Gallows*, by Pieter Bruegel the Elder. As this mesmerizing painting shows, M. C. Escher was not the first one to use impossible objects in art. *Wikimedia Commons*. The work of art depicted in this image and the reproduction thereof are in the public domain worldwide. The reproduction is part of a collection of reproductions compiled by The Yorck Project. The compilation copyright is held by Zenodot Verlagsgesellschaft mbH and licensed under the GNU Free Documentation License.

that, on that occasion, we should perhaps have less confidence in our intuition.

To argue the case, I will use an extended version of Euler's Theorem. I will try to formulate my proof in a language of elementary mechanics which would not be out of place in Euler's time.

Euler's Theorem (Section 2.3) describes the possible motions of a solid body around a fixed point: these are rotations around appropriate axes. A free motion of a body can be viewed as a composition of a rotation and a parallel translation. Rotations have three degrees of freedom (two of them define the axis and the third one the angle of rotation). Parallel translations have three degrees of freedom, corresponding to the three coordinates in space. Therefore movements of a solid body have six degrees of freedom.

Daina Taimina, aged 12

When moving one piece of glass over the other (we assume that the latter is fixed to the workbench), we have three degrees of freedom: movements forward–backward, left–right and rotations. A "thorough" grinding means that all three degrees of freedom are used. It will be convenient for us to restrict ourselves to cautious and tiny "local" movements. A composition of two small (one may wish to say "infinitesimal") grinding movements is a (perhaps slightly longer) grinding movement, and the reverse movement is also a grinding movement. We shall say that grinding movements form a (local) subgroup in the group of motions of the solid body.

The following theorem is the key observation.

> In the group of motions of a solid body, only three types of subgroups have exactly three degrees of freedom:
> - the group of all parallel translations of the three-dimensional space (which we cannot use since we have to keep the two pieces of glass together);
> - the group of all rotations around a fixed point—it produces a spherical mirror; and
> - the group of all movements which preserve a plane—it consists of compositions of parallel translations in the plane and rotations around an axis perpendicular to the plane. This group produces a flat mirror.

Fig. 4.5. A spherical stone vase, Ancient Egypt. © Petrie Museum of Egyptian Archaeology, University College London; item UC 41616, reproduced with permission. The vase was made by polishing a piece of stone. Its spherical shape is a consequence of the description of subgroups in the Lie group of isometries of the three-dimensional Euclidean space.

Most likely, the statement of the theorem is not immediately self-evident to most readers. And this was my intention: I wished to demonstrate a threshold at which our sensorimotor intuition starts to falter. The statement of the theorem was not self-evident to me since, to tell the truth, I cheated and used some algebra to find the formulation and a proof of the theorem—I could not trust my sensorimotor intuition. I feel that while the concept of "degrees of freedom" is sufficiently intuitive, the concept of a group of movements is much harder. First of all, we struggle to predict the result of the consecutive execution of two movements.

We can easily visualize a collection of objects if we have *seen* them—one man or a crowd, one flower or the whole garden in blossom. But it is very difficult for a human to form a mental image of a multitude of movements of his or her hand and treat this multitude as a single entity.

But the theorem would perhaps be self-evident to mathematicians who were developing theoretical mechanics during the 18th and 19th centuries—they frequently solved problems about real mechanisms made from axles, levers, etc., and therefore were likely to have a good intuition about movements in space, since they *have seen* these mechanisms working.

The theorem is based on a range of ideas which led to the creation of the theory of Lie groups and Lie algebras (or *infinitesimal groups*, in the terminology of the 19th century). When one reads old papers on Lie groups, it is striking how alien was the concept of a set to their authors—they were quite happy to use continua of various dimensions.

4.6 Mental arithmetic and the method of Radzivilovsky

> *If your experiment needs statistics,*
> *you ought to have done a better experiment.*
> Ernest Rutherford

The emphasis on the importance of connections between various neural systems responsible for mental arithmetic led a prominent neuropsychologist, Brian Butterworth [160], to propose not teaching arithmetic until the student was of the age of 11. In Butterworth's view, the traditional way of teaching over-exploits verbal counting to the detriment of the development of the "approximate" arithmetic of magnitude.

Butterworth was apparently inspired by the success of Louis Benezet's experiments (in the 1920s in the USA); indeed, Benezet did not formally teach children arithmetic [440, 116]. Instead, Benezet's method involved a great deal of estimation activities in a real life setting.

I do not see the reason why estimation activities and "approximate" arithmetic should be set up against "exact" arithmetic. From my point of view, the key to teaching arithmetic is the synthesis of all three.

I know a brilliant and idiosyncratic mathematics teacher, Vladimir Radzivilovsky, whose methods included asking (very) young children to guess the weight of and then weigh (in grams) every household item which they could fit on a scale. Radzivilovsky also asked children to estimate temperature (by touching water, say) and then to compare their feelings with a measurement by a thermometer. On the other hand, exact symbolic arithmetic was also drilled into the children with a persistence and rigor unlikely to be found in any school in the world: Radzivilovsky's method involves forward and backward speed counting by ones, twos, threes, etc., with the performance being timed. These counting exercises led to the times tables and lists of prime numbers being *composed by each child* and only afterwards memorized for subsequent reuse.

It appears that Radzivilovsky systematically built bridges between various mental presentations of order and number in his pupils' heads. If these lines are read by a teacher or an education theorist, I have to disappoint (or comfort) him or her by explaining that Radzivilovsky worked privately outside the school system and taught only individuals or very small groups. This allowed him to rigorously stick to the key element of his system: he never moved to the next stage until his pupil had reached absolute, automatic perfection in handling simpler mathematical objects and concepts.

Vladimir
Radzivilovsky,
aged 7

Radzivilovsky's success is measured by the fact that, among his former students, he can name *dozens* of professional mathematicians, physicists, and computer scientists, or, if they are still young, finalists of International Mathematical Competitions. Radzivilovsky believes that teaching is an art, not a science. Moreover, teaching, in his opinion, is a performance art and therefore he, unfortunately, does not see the point in putting his ideas in writing.

4.7 Not-so-simple arithmetic: "named" numbers

> *Toutes les grandes personnes ont d'abord été des enfants.*
> *(Mais peu d'entre elles s'en souviennent.)*
> Antoine de Saint-Exupéry, *Le Petit Prince*

I pay so much attention to basic arithmetic for two reasons. First, cognitive aspects of arithmetic have been intensively studied (possibly, for good reason) by neurophysiologists who accumulated many interesting observations. The other reason is that I myself, as a child, had serious psychological difficulty with arithmetic. I will go into this more in Section 5.2; here, I discuss the issue that many readers would consider trivial: "named" numbers, like two apples and three people. (In Britain and the USA, the term "numbers with 'units'" is sometimes used for "named numbers".)

After I was told by my teacher that I had to be careful with *"named"* numbers and not to *add* apples and people, I remember asking her why in that case we can *divide* apples by people:

$$10 \text{ apples} \div 5 \text{ people} = 2 \text{ apples}.$$

This is even worse: when we distribute 10 apples giving 2 apples to a person, we have

$$10 \text{ apples} \div 2 \text{ apples} = 5 \text{ people}.$$

Where do "people" on the right-hand side of the equation come from? Why does "people" appear and not, say, "kids"? There were no "people" on the left-hand side of the operation! How do numbers on the left-hand side know the name of the number on the right-hand side?

I did not get a satisfactory answer from my teacher and lived for a couple of years with this torment until we started to solve problems about speed and distance. Only then did I realize that the correct naming of the numbers in these examples should possibly be

$$10 \text{ apples} \div 5 \text{ people} = 2 \, \frac{\text{apples}}{\text{people}}$$

and

$$10 \text{ apples} \div 2 \, \frac{\text{apples}}{\text{people}} = 5 \text{ people}.$$

But that discovery made me even more uncomfortable: I knew that

$$\text{speed} = \frac{\text{distance}}{\text{time}},$$

but what was the nature of the strange substance,

$$\frac{\text{apples}}{\text{people}}?$$

Now, forty years later, I know the answer. It is commonplace wisdom that the development of mathematical skills in a student goes alongside the gradual expansion of the realm of numbers with which he or she works, from natural numbers to integers, then to rational, real, complex numbers:[9]

$$\mathbb{N} \subset \mathbb{Z} \subset \mathbb{Q} \subset \mathbb{R} \subset \mathbb{C}.$$

What is missing from this natural hierarchy is that already at the level of elementary school arithmetic children are working in a much more sophisticated structure, a graded ring

$$\mathbb{Q}[x_1, x_1^{-1}, \ldots, x_n, x_n^{-1}]$$

of Laurent polynomials in n variables over $\mathbb{Q}$, where the symbols

$$x_1, \ldots, x_n$$

stand for the names of objects involved in the calculation: apples, persons, etc.

The ring $\mathbb{Q}[x_1, x_1^{-1}, \ldots, x_n, x_n^{-1}]$ is just the set of all polynomials (called *Laurent polynomials*) with rational coefficients in variables x_i *and* x_i^{-1}; for example,

$$(3/4) + x_1 x_2 x_3^{-2} + (1/2)x_1^{-1}$$

is such a polynomial. The operations of addition and multiplication are defined as usual. Monomials can be inverted as well:

$$(x_1 x_2^{-1})^{-1} = x_1^{-1} x_2.$$

Usually, only monomials are interpreted as having physical (or real life) meaning. But the addition of heterogeneous quantities still makes sense and is done componentwise: if you have a lunch bag with (2 apples + 1 orange) and another bag with (1 apple + 1 orange), together they make

(2 apples+1 orange)+(1 apple +1 orange) = (3 apples+2 oranges).

Notice that this gives a very intuitive and straightforward approach to vectors and vector algebra. In Section 12.6.4 I mention that this "lunch bag" approach to vectors allows a very natural introduction to duality and the vector/covector notation in tensor algebra: the total cost of a purchase of amounts g_1, g_2, g_3 of some goods at prices p^1, p^2, p^3 is $\sum g_i p^i$. In particular, this allows us to see that the quantities g_i and p_i could be of completely different natures.

Of course, there is no need to tell all that abstract nonsense about Laurent polynomials to kids;[10] but it would not harm the teachers of mathematics to be prepared to answer a child's question about whether it is legitimate to work with the named quantity

$$\frac{\text{apples}}{\text{people}}.$$

The answer is yes, of course; the quantity is conveniently called "apples per person". However, people sometimes give new names to new quantities, sometimes not. For example,

$$\frac{\text{money}}{\text{people} \times \text{time}}$$

is often called *wage*.[11]

The extreme form of "named numbers" is numerals used for counting specific types of objects (most likely, they historically precede the emergence of the universal number system as we know it). In England, a popular slander about Yorkshiremen is that they use special numerals for counting sheep. Judging by the Lakeland Dialect Society website [449], local people proudly admit to sticking to the old ways. In Wensleydale, for example, the first ten sheep numerals are said to be

> yan, tean, tither, mither, pip, teaser, leaser, catra, horna, dick.

We shall return to sheep numerals later; see page 250.

If we turn to more modern times, it is entertaining to compare sheep numerals with Richard Feynman's joke [427]:

> You see, the chemists have a complicated way of counting: instead of saying "one, two, three, four, five protons", they say, "hydrogen, helium, lithium, beryllium, boron."

Physicists love to work in the Laurent polynomial ring

$$\mathbb{R}[\text{length}^{\pm 1}, \text{time}^{\pm 1}, \text{mass}^{\pm 1}]$$

because they love to measure all physical quantities in combinations (called "dimensions") of the three basic units: for length, time, and mass. But then even this ring becomes too small since physicists have to use fractional powers of basic units. For example, velocity has dimensions length/time, while electric charge can be meaningfully treated as having dimensions[12]

$$\frac{\text{mass}^{1/2}\text{length}^{3/2}}{\text{time}}.$$

Well, we should not be too hard on physicists for their excessive reductionism—we all know people who measure everything on just two basic scales: time and money. As we all know, mass can be easily converted into money: when I ask the salesgirl at the deli counter in a local (British) supermarket to slice a pound of ham, she usually asks me back: "Pound in weight or pound in money?" And, of course, we also know that "time is money".

It pays to be attentive to the dimensions of the quantities involved in a physical formula: very frequently the balance of dimensions of the left- and right-hand sides of the formula suggests the shape of the formula. Physicists call such a way of reasoning *dimensional analysis*. In Section 8.4 I give an example of the application of dimensional analysis to the deduction of a stunningly beautiful formula: Kolmogorov's celebrated "5/3" law for the energy spectrum of turbulence.

Notes

[1]DIRICHLET PRINCIPLE. The name *Dirichlet Principle* can possibly be traced back to European, primarily German tradition; it is used by Freudenthal [113]. It would be interesting to read a well-documented history of the Pigeonhole Principle (who invented the name?). Meanwhile, the referee of the present book wrote in his comments: "I'm pretty sure the name is because Dirichlet used a pigeonhole argument to solve Pell's equation. He modestly omitted his argument from his *Vorlesungen über Zahlentheorie*, but Dedekind included it as an appendix in the 1871 edition. See the English translation of Dirichlet, *Lectures on Number Theory* (History of Mathematics, AMS, 1999), p. 258."

[2]SUBITIZING. One has to distinguished between subitizing and recognition of a geometric pattern. David Pierce sent me a very nice example:

> Old playing cards did not have their values printed in the corners; one knew their values only by looking at the pips. But with experience, one would not *count* the pips; one would just know that the card

> had the value 10.

In this example, one can learn to recognize a specific pattern formed by pips and correlate it with the number 10; however, one cannot recognize 10 randomly positioned pips without counting. I would be most happy to have a look at the results of experimental studies—if there were any—of how subitizing of a group of objects is affected by symmetries in the position of objects.

Gregory Cherlin commented that perhaps the Pigeonhole Principle plays a role in Sudoku, but more in a counting form: e.g., if three numbers have to fit into three spaces and only one per space, then there are not any other numbers there.

[3]MUSICAL SCALE. Most likely, the seven notes of the musical scale have nothing to do with subitizing. There is a classical mathematical explanation of the structure of musical scales: see, for example, Benson [9]. John Stillwell commented on this that Newton enumerated 7 colors of the rainbow because he believed they should correspond to the notes of the scale. See Newton's *Optics*, Book One, Part II, Exper. 7 (p. 128 in the Dover edition [74]).

[4]Or compare with Lewis Carroll:

> "Can you do Addition?" the White Queen asked. "What's one and
> one and one and one and one and one and one and one and one
> and one?"
> "I don't know," said Alice. "I lost count."
> "She ca'n't do Addition," the Red Queen interrupted.

(Lewis Carroll, *Through the Looking-Glass, and What Alice Found There*,
1875)

[5]Notice the remarkable mathematical concept: a *winning strategy*. It is
relatively modern; it is not taught at schools; it somehow has crept into
mass culture and is known to most school-children. Whether they are prepared to deal with it as a *mathematical* entity is a different issue; see more
on this in Section 6.1.

[6]I owe this comment on ordinals and cardinals to David Pierce.

[7]The interpretation of Euclid's definition of a straight line as a line of
sight was suggested to me by David Pierce and supported by Alexander
Jones. See a detailed discussion of "straightness" in [272, Chapter 1] by
David Henderson and Daina Taimina.

[8]For the sake of rigor, a "body" is a compact subset X in $\mathbb{R}^n$ with the
connected complement $\mathbb{R}^n \smallsetminus X$.

[9]THE HIERARCHY OF NUMBER SYSTEMS. Let us return for a minute to
the traditional hierarchy of number systems

$$\mathbb{N} \subset \mathbb{Z} \subset \mathbb{Q} \subset \mathbb{R} \subset \mathbb{C}.$$

At the formal level, the step from $\mathbb{Q}$ to $\mathbb{R}$ is highly non-trivial; the steps
from $\mathbb{Z}$ to $\mathbb{Q}$ and from $\mathbb{R}$ to $\mathbb{C}$ are somewhat simpler but require a considerable level of abstract algebra. At the secondary school level, details are
usually omitted, and rightly so. As a result, many natural questions which
an inquisitive child is very likely to ask remain unanswered. As a child,
I asked my elementary school teacher why it was forbidden to divide by
zero—and was ridiculed for my question.

I believe that, at that time, I would be quite satisfied to hear an explanation along the following lines:

> "When we multiply 2 by 3, the number 2 is hidden inside of the
> result, 6, and we can get it back by dividing 6 by 3. Multiplication
> by 0 destroys the number; whatever we multiply by 0, we get 0;
> no memory of the old number is left, and it cannot be recovered.
> This is the reason why we do not divide by 0—it would not help us
> anyway, even if we did so."

Gregory Cherlin told me that, in his undergraduate class on Foundations of Mathematics he had a third-year undergraduate ask why division
by zero was impossible. What was interesting is her explanation that she
had been wanting to ask this question since 5th grade but only now had
found both a context and a level where she thought it was likely that she
could get an answer—that was because the rationals were built as equivalence classes of pairs of integers.

[10]CARRIES AND GROUP COHOMOLOGY. The deceptive simplicity of the
elementary school arithmetic is especially transparent when we take a
closer look at *carries* in the addition of decimals.

In Molièr's *Le Bourgeois Gentilhomme*, Monsieur Jourdain was surprised to learn that he had been speaking prose all his life. I was recently reminded that, starting from my elementary school and then all my life, I was calculating 2-cocycles; I thank Mikael Johansson who brought this fact to my attention.

Indeed, a carry in elementary arithmetic, a digit that is transferred from one column of digits to another column of more significant digits during addition of two decimals, is defined by the rule

$$c(a, b) = \begin{cases} 1 & \text{if } a + b > 9, \\ 0 & \text{otherwise.} \end{cases}$$

One can easily check that this is a 2-cocycle from $\mathbb{Z}/10\mathbb{Z}$ to $\mathbb{Z}$ and that it is responsible for the extension of additive groups

$$0 \longrightarrow 10\mathbb{Z} \longrightarrow \mathbb{Z} \longrightarrow \mathbb{Z}/10\mathbb{Z} \longrightarrow 0.$$

Of course, what else could it be?

[11]When, as a child, I was nursing my doubts about the way "named" numbers were taught at my school, I was unaware that I had an ally in François Viète, who in 1591 clearly wrote in his *Introduction to the Analytic Art*:

If one magnitude is divided by another, [the quotient] is heterogeneous to the former . . .

He added:

Much of the fogginess and obscurity of the old analysts is due to their not paying attention to these [rules]. [85, p. 16]

[12]THE DIMENSIONS OF ELECTRIC CHARGE. If we choose our units in such a way that the permittivity ϵ_0 of free space is dimensionless, then from Coulomb's law

$$F = \frac{1}{4\pi\epsilon_0} \frac{q_1 q_2}{r^2}$$

applied to two equal charges $q_1 = q_2 = q$, we see that q^2/r^2 has the dimensions of force.

5

Infinity and Beyond

There is no progression into infinity; why not?
because the human intellect must have some foundation?
because it is accustomed to this belief?
because it cannot imagine anything beyond its own limits?
As if, indeed, it followed, that if I do not comprehend infinity,
therefore there is no infinity.

De Tribus Impostiribus [453]

In this chapter I discuss how we interiorize infinity; I do not try to address the extremely hard philosophical issues immediately arising as soon as we touch one of the thorniest subjects in the methodology of mathematics. I talk only about the intuition of infinity as it manifests itself in everyday (and not very advanced) mathematical work. I try to avoid any technicalities and stay strictly within naïve set theory.

First, I try to show, using the material from the running example of the book, namely, symmetry and Coxeter Theory, that we are quite inept at visualizing infinity. However, we are surprisingly good at dealing with the potential infinity of words and numbers.

I discuss potential infinity in Sections 5.3 and conclude the chapter by revisiting the geometric intuition of infinity. I prefer to postpone the discussion of *actual* infinity until later in the book, Section 6.2.

5.1 Some visual images of infinity

All things [are] full of labour; man cannot utter [it]:
The eye is not satisfied with seeing,
nor the ear filled with hearing.

Ecclesiastes 1:8

Fig. 5.1. A tessellation of the sphere by the mirrors of symmetry of an icosahedron (or, what is equivalent, by mirrors of reflections belonging to the finite reflection group H_3).

I return to the running example of the book, symmetry and the theory of Coxeter groups. As Paul Moszkowski demonstrated in his paper [388], it is perfectly possible to develop the theory and classification of finite Coxeter groups entirely in terms of words, without any recourse to geometry. Almost all existing treatments of the theory rely on geometry to some degree, and I personally prefer an entirely geometric treatment. However, when you start thinking about *infinite* Coxeter groups, the use of word-based methods is much harder to avoid.

Indeed, the system of all mirrors of symmetry belonging to a finite reflection group can be visualized with the help of pictures like Figure 5.1. A similar picture for *hyperbolic reflection groups* has to be drawn on the hyperbolic plane and results in images like Figure 5.2.

For a novice, the mess at the limiting circle of the hyperbolic plane might be disturbing. This makes it much more difficult to directly use hyperbolic tessellations in "pictorial" proofs: either you have to have a well-developed intuition about hyperbolic geometry, or supplant it by algebraic manipulations. A page in a book is perceived by us as a fragment of the Euclidean plane; mapping from the sphere or the hyperbolic plane to the Euclidean plane necessarily distorts the actual distances and sizes of the objects.

Coxeter understood the methodological difficulties arising from the explanation of the geometry of hyperbolic reflection groups. In the 1960s he made several films popularizing his ideas (see their discussion in [29]). In one of them, he addressed the audience while holding in his hands a diagram on a piece of paper, similar to Figure 5.2, and a big glass sphere, painted like a Christmas tree dec-

Fig. 5.2. A tessellation of the hyperbolic plane formed by mirrors of symmetries belonging to the hyperbolic reflection group
$$\langle a, b, c \mid a^2 = b^2 = c^2 = (ab)^6 = (bc)^4 = (ac)^2 = 1 \rangle.$$
Using the "palindrome" terminology of Section 3.4, it is a Coxeter language with the alphabet $\{a, b, c\}$ and basic equivalencies
$$ababab = bababa, \quad bcbc = cbcb, \quad \text{and } ac = ca.$$

oration, with the pattern of Figure 5.1. Coxeter explained that the perceived decrease in the size of triangles on the hyperbolic plane closer to the limiting circle is an illusion, similar to the perspective deformation of the tessellation on the sphere. The limiting circle could be usefully thought of as the "horizon" of the hyperbolic plane. This interpretation was a pedagogical masterstroke, because, as we shall see in Section 5.5, it referred to the built-in mechanisms of the "recalculation" of perspective deformation in our brains.

However, although we can easily believe that the actual sizes of the triangles in Figure 5.2 are the same, it is much harder to accept that the pattern of fitting triangles together is the same at the blurred "horizon" as in the center of the diagram. Just compare Figure 5.2 with the mirror tessellations of the Euclidean plane, shown flat out, without perspective: here we have no doubt that exactly the same pattern goes all through the entire plane (Figure 5.3).

When we deal with infinity, we want to be in control; as mathematicians, we do not want to just stare at it in awe. Losing control, as in a bad dream, can be a powerful aesthetic experience

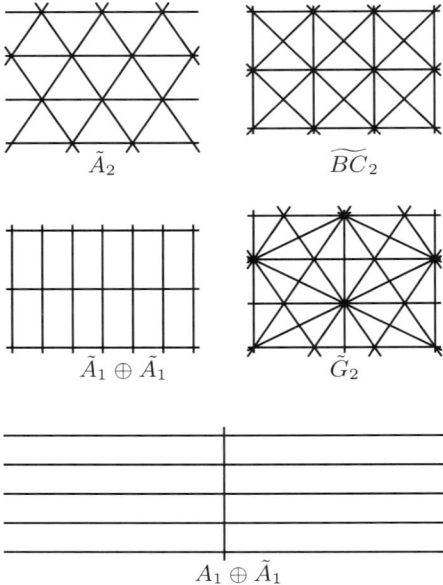

Fig. 5.3. *Tessellations of the Euclidean plane by mirrors of symmetry belonging to Euclidean reflection groups with their traditional notation: congruent equilateral triangles $\tilde{A}_2$; isosceles right triangles $\widetilde{BC}_2$; rectangles $\tilde{A}_1 \times \tilde{A}_1$; triangles with the angles $\pi/2$, $\pi/3$, $\pi/6$ $\tilde{G}_2$; and infinite half stripes $A_1 \times \tilde{A}_1$. Drawing by Anna Borovik.*

and is fully exploited as such in Escher's famous engravings from the *Circle Limit* series[1]—or in the famous sequence in *2001: Space Odyssey*. But it is not what we want from our mental images of infinity.

So far all of my examples of "bad" infinity were two-dimensional, and, perhaps, this was a possible source of confusion. As we shall soon see, we are quite happy to deal with the one-dimensional infinity of words and numerals.

5.2 From here to infinity

> *Straszne, ze wiecznosc sklada sie z okresow sprawozdawczych.*
> What really scares me is that eternity is made of deadlines.
> Stanislaw Jerzy Lec

> *I have seen the future and it is just like the present, only longer.*
> Kehlog Albran

Fig. 5.4. A pattern of angels and devils based on M. C. Escher's *Circle Limit IV* and the hyperbolic tessellation $\{4, 5\}$. Rendition by Douglas Dunham [28], reproduced with his kind permission.

Mathematics, as we know it, is possible only because the language processing system in our brains provides us with powerful mental tools for dealing with potentially infinite sequences of mental objects. At the interface with mathematics, we can identify two reasons for that.

The first one is that our natural language is potentially infinite. We discover infinity not when we look at the stars in the night sky, not when we see railway tracks merging at the horizon; mathematical infinity is usually first encountered when a realization dawns on us that the purely linguistic exercise of reciting numerals produced by certain fixed and entirely linguistic rules will apparently never end. I remember how excited and shocked my son was, then four years old, when he found himself on this endless numerological treadmill. Recently, I relived these memories when I watched the charming French documentary *Etre et Avoir* [426], where in a similar scene, the teacher, Monsieur Lopez, nudged the puzzled and somewhat sceptical child, little Jojo, into counting on and on.

The second reason is that the language processing modules of our brains are built to deal comfortably with potentially endless language inputs:

. . . nor the ear filled with hearing. . . (Ecclesiastes 1:8)

Fig. 5.5. The underlying tessellation of M. C. Escher's *Circle Limit IV* is the hyperbolic tessellation of type $\{6, 4\}$: it is made up of regular hexagons (6-gons), with 4 hexagons meeting at each vertex. In this picture, if one cuts every hexagon by diagonals into 6 triangles thus separating devils from angels, one gets a hyperbolic mirror system corresponding to the set of palindromes in the Coxeter language with the alphabet $\{a, b, c\}$ and basic equivalences

$$aba \equiv bab, \quad acac \equiv caca, \quad bcbc \equiv cbcb.$$

Equivalence classes of synonymous words of the Coxeter language can be used to label the angels and devils, giving us control over the sprawling chaos.

The tessellation of type $\{4, 5\}$ in Figure 5.4 consists of hyperbolic squares (or regular 4-gons), with 5 squares meeting at each vertex.

Rendition by Douglas Dunham [28], reproduced with his kind permission.

Even more importantly, language processing is predictive, in that we subconsciously try to guess the next word. This is why all kinds of limits at infinity, completions and compactifications are easier to comprehend when they are represented by sequences or words. The examples are abundant; we have all encountered infinite decimal expansions like

$$\pi = 3.1415926\ldots$$

and even trickier ones like

$$1 = 0.9999\ldots\,.$$

If you compactify the integers in the 2-adic topology, you come to 2-adic integers which can be conveniently represented by sequences of binary digits infinite to the left:

$$-\frac{1}{3} = \frac{1}{1-4}$$
$$= 1 + 4 + 4^2 + 4^3 + \cdots \qquad \text{(by the formula for a geometric progression)}$$
$$= 1 + 100 + 10000 + \cdots \qquad \text{(in binary notation)}$$
$$= \ldots 1010101.$$

Or you may wish to use continued fractions to represent the golden section as the limit of an iterative algorithm:

$$\frac{1+\sqrt{5}}{2} = 1 + \cfrac{1}{1 + \cfrac{1}{1 + \cfrac{1}{1 + \cfrac{1}{1 + \cdots}}}}$$

My own first clash with infinity happened at the level of grouping objects in counting, which is, I now suspect, something very similar to bracketing or parsing. I remember my extreme discomfort when, as a child, I was taught division. I had no bad feelings about dividing 10 apples among 5 people, but I somehow felt that the problem of deciding how many people would get apples if each was given 2 apples from the total of 10, was completely different. (My childhood experience is confirmed by experimental studies; see Bryant and Squire [158].)

Mathematics is about the reproducibility of our mental constructions, hence about control.

In the first problem you have a fixed data set: 10 apples and 5 people, and you can easily visualize giving apples to the people, in rounds, one apple to a person at a time, until no apples are left. But an attempt to visualize the second problem in a similar way, as an orderly distribution of apples to a queue of people, two apples to each person, necessitates dealing with a potentially unlimited number of recipients. In horror I saw an endless line of poor wretches, each stretching out his hand, begging for his two apples. This was visualization gone astray. I was not in control of the queue! But reciting numbers, like chants, while counting *pairs* of apples, had a soothing, comforting influence on me and restored my shattered confidence in arithmetic.[2]

As soon as I started to consult the literature, I discovered that some of my observations had been made before. In this particular case, Frank Smith had already used the expression "mathematical chant" in his book *The Glass Wall* [132].

The mother of all iterative processes is counting, and the potentially infinite set of natural numbers is the mother of all potential infinities.

As I have said on many occasions in this book, mathematics is about the reproducibility of our mental constructions, hence about *control*. We want to control the mathematical objects we create, we want to be able to deal with them as with real life objects, we want to perform *actions*.

I remember that, at the age of three or four, I (like many children of that age) had a so-called eidetic imagination: I could close my eyes and see things at will, with all their details and colors, almost indistinguishable from real things in real life. I could see a car and I could open the door of this car—and it opened as a real door in a real car. Later this disappeared: my brain learned to save resources and compress the images (with loss of data, as always happens in compression) into more manageable, compact, and easy-to-store formats.

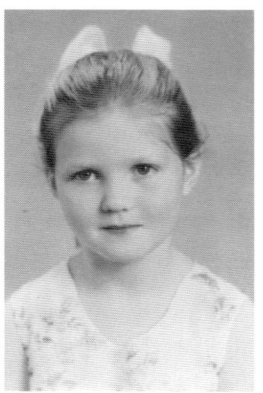

Anna Borovik,
nee Vvedenskaya,
aged 10

Anna, my wife, told me that her last eidetic episode was at age of 10: her parents sent her to bed and did not permit to finish the book she was reading. Anna glanced over the last two pages without having time to read a single word. In her dream that night, she read the two pages. In the morning she checked the book: her reading in sleep was correct, including a word she had never encountered before, but guessed its meaning.[3]

The nature of eidetism still remains a mystery. I was alarmed to read the following in Lorna Selfe [228, p. 112]:

Eidetic imagery was once thought to be a normal stage of development in all children and therefore to be related to other facts of early cognitive development, such as sensory rather than verbal modes of encoding experience, and concrete rather than abstract modes of thought. However, recently the question has been raised as whether eidetism is a normal phenomenon with adaptive significance or whether it is essentially maladaptive and a direct manifestation of brain pathology.

This was written in 1977, and I would like to know whether the medical assessment of eidetism has changed.

When learning or doing mathematics, we quite frequently have to create mental images of mathematical objects with eidetic qualities as close to that of the images of real objects as possible. (If we are duped, as a result, into the belief that mathematical objects exist in some ideal or dream world, this happens only because we want to be duped.) Sometimes such images are easy to create (for me the concept of "triangle" has the same eidetic intensity as that of "chair", say), and sometimes it is more difficult; using a computer technology metaphor, we are frequently reduced to moving the icons for graphic files around the screen instead of seeing the contents of the files.

When it comes to our intuitive images of infinity, how do we control them? How does this affect our handling of mathematical infinity?

I try to answer these questions in the next section.

5.3 *The Sand Reckoner* and potential infinity

> *There is no smallest among the small*
> *and no largest among the large;*
> *but always something still smaller*
> *and something still larger.*
> Anaxagoras

We have to distinguish between the *potential infinity* of an iterative process which could be continued on and on and the *actual* infinity of the output of this process being imagined as completed, *encapsulated* and made into an object.[4]

The mother of all iterative processes is counting, and the potentially infinite set of natural numbers is the mother of all potential infinities. A word of warning is needed: we have to distinguish between iteration and repetition. The sun rising in the morning is, of course, the mother of all repetitive processes; but the sun is the same, yesterday, today, and tomorrow, while numbers are all *different*. However, for the bulk of its history, mathematics was a branch of astronomy; without doubt, the extreme precision of repetition of many astronomic phenomena very much influenced the development of the culture of mathematical rigor. Let us perform a small thought experiment: imagine that atmospheric conditions on Earth were, for the last 5,000 years, slightly different: a light haze obscured the stars in the night sky (without adversely affecting the climate and conditions for the development of agriculture, etc.). How would mathematics develop? Would it ever reach the stage beyond basic arithmetic and purely procedural geometry, without proofs?

The crucial importance of the potential infinity of natural numbers *as a mathematical problem* was realized early on, and Archimedes wrote a fascinating book, *The Sand Reckoner*, bringing the potential infinity of natural numbers home to his contemporaries. To that end, he developed elaborate terminology to describe the number of grains of sand in bigger and bigger spheres, with the radius reaching the Sun and growing further. We have to take notice that the potential infinity of natural numbers required demonstration; the book was deemed, over the centuries, important enough to be saved and copied, so that the text survived. One of the happiest moments in my teaching life involved passing around in a calculus class a translation of *The Sand Reckoner*. *The Sand Reckoner* was probably the first book ever in the genre of "popular mathematics"—and remains a masterpiece even if judged by modern standards.

The opening line of the book is wonderful:

> Some people believe, King Gelon, that the number of sand is infinite in multitude.

Archimedes needs just two more sentences to complete the set-up of the problem:

> I mean not only of the sand in Syracuse and the rest of Sicily, but also of the sand in the whole inhabited land as well as the uninhabited. There are some who do not suppose that it is infinite, and yet that there is no number that has been named which is so large as to exceed its multitude.

Thus the first paragraph of the book already contains the key idea of the solution: we need names. Regarding indefinitely developing processes, as long as we can give names for some of their intermediate instances, which are spread, like milestones, all over the process, we feel comfortable, we are not afraid that we may run out of names.

Archimedes describes an iterative process of building bigger and bigger masses of sand; however, he feels that he is in firm control of the potentially infinite process because, in modern terminology:

- The individual instances produced in the process are related by linear order (magnitude).
- Some instances (objects) are measured (or "counted") by natural numbers, and this measure is compatible with the order: bigger objects have bigger measures.
- The linear order satisfies what is now known as *Archimedes' Axiom*: for every object there is a bigger measured object.[5]

These are the most basic, "atomic" types of potential infinity, and they are the easiest to handle, for reasons that we have already discussed in Section 4.2: the mental images of potential infinity are built on the basis of the pre-existing hardwired structures of our minds: order and numerals. For example, we see the potential infinity of time through the potential infinity of the calendar.

Nowadays the acceptance of the potential infinity of natural numbers is just part of common, everyday culture, and children (like little Jojo) absorb it at a very early age. In any case I never encountered a student who would question the existence of potential infinity; the issue is whether our students are able to control its simplest manifestations; for example, can they compute the 100-th term of the sequence

$$1, 3, 5, 7, 9, 11, \ldots?$$

Nowadays the acceptance of the potential infinity of natural numbers is just part of common, everyday culture.

I suspect that, in a young child, a healthy scepticism about the possibility of counting indefinitely might be more a sign of potential mathematical abilities (because she may need to check for herself that numerals, indeed, do not get out of control) than the readiness to accept, already at the 20th term, that the dull routine will drag on forever.

We frequently forget, however, that the potential infinity of natural numbers is already an abstraction, the result of encapsulating the necessarily finite process of counting as part of an idealized infinite process. In common teaching practice, the intermediate steps in the abstraction are skipped or taken for granted. To illustrate our carelessness, consider the following thought experiment (taken from the book *Mathematical Aquarium* [292] by Victor Ufnarovski where it is formulated as a "competition style" problem).

In a young child, a healthy scepticism about the possibility of counting indefinitely might be more a sign of potential mathematical abilities than the readiness to accept, already at the 20th term, that the dull routine will drag on forever.

Assume that we are given an extremely reliable computer with an eternal source of electric supply. The computer is programmed to print, via an external printer with an unlimited supply of paper and cartridges, consecutive natural numbers:

$$1, 2, 3, 4, 5, 6, 7, 8, 9, 10, 11 \ldots.$$

(However, it cannot read its output.) Prove that the computer will sooner or later fail.

Indeed, it will fail for reasons of its intrinsic limitations: the computer has only a finite number of internal states (since it has only finitely many memory cells, for example); printing a new number requires a change in some of these states. Since the computer is working indefinitely, some of its states reoccur, say, at moments of time T_1 and T_2; but then it will print, from time T_1 on and from time T_2 on, the same sequence of numbers, which means that it cannot do its job properly. (Notice that this argument can be viewed as an application of the Pigeonhole Principle, Section 4.1.) In short, the computer will fail because of the eventual overflow of memory. Of course, the answer would be different if the computer could read its own output and use the paper tape as an external memory device.

Our brains are finite state machines, and we cannot count forever, not because we are mortal but because our brains are finite.

Of course, you may wish to try to circumvent the problem by inventing ever more elaborate and compact abbreviations, but all successful ones will require recursion (reference to abbreviated names for previous numbers); if recursion is accepted, the counting becomes reduced to repeating, again and again, "the previous number plus one, the previous number plus one, the previous number plus one", which, we have to admit, *is* the most compact form of the description of the natural numbers. This is, by the way, why a famous programming language is called C^{++}: it was developed from the language C, and in both languages the operation of incrementing, in recursive procedures, of a value n by 1 was deemed to deserve special notation: $n{+}{+}$.

Still, we have a built-in facility for thinking about processes and sequences of events, and as long as potential infinity appears as a linear process or a sequence of events (and is interspersed with a "counted" subsequence labelled by numerals), we usually have no trouble in interiorizing it. The sprawling infinity of hyperbolic tessellations (Figures 5.2 and 5.4) is much less intuitive because we soon lose control of the intermediate steps: even if you know that the pattern of adjacency is the same, it is somehow hard to believe.

5.4 Achilles and the Tortoise

> *Quoth the raven, 'Nevermore.'*
> Edgar Allan Poe

Of all the paradoxes of infinity, Zeno's "Achilles and the Tortoise" paradox is one of the oldest. I borrow its description from WIKIPEDIA:

> "In a race, the quickest runner can never overtake the slowest, since the pursuer must first reach the point whence the pursued started, so that the slower must always hold a lead" (Aristotle, *Physics* VI:9, 239b15). In the paradox of Achilles and the Tortoise, we imagine the Greek hero Achilles in a footrace with the plodding reptile. Because

he is so fast a runner, Achilles graciously allows the tortoise a head start of a hundred feet. If we suppose that each racer starts running at some constant speed (one very fast and one very slow), then after some finite time, Achilles will have run a hundred feet, bringing him to the tortoise's starting point; during this time, the tortoise has "run" a (much shorter) distance, say one foot. It will then take Achilles some further period of time to run that distance, during which the tortoise will advance farther; and then another period of time to reach this third point, while the tortoise moves ahead. Thus, whenever Achilles reaches somewhere the tortoise has been, he still has farther to go. Therefore, Zeno says, swift Achilles can never overtake the tortoise. Thus, while common sense and common experience would hold that one runner can catch another, according to the above argument, he cannot; this is the paradox.

In view of our discussion in the previous section, the most natural approach to the paradox is complexity-theoretic. Indeed, in the description of the race between Achilles and the Tortoise, we have two different timescales: the one in which the motion of Achilles and the Tortoise takes place and another one in which we discuss their motion, repeating again and again the words

> "it will then take Achilles some further period of time to run that distance, during which the tortoise will advance farther".

Clearly, each utterance takes time bounded from below by a non-zero constant; therefore the sum of the lengths of our utterances diverges. However, our personal time flow has no relevance to the physical time of the motion!

Since Zeno's paradox is not about mathematics as such, but about its relations to the real world and about our perception of time, a complexity-theoretic approach to its solution is well justified. The validity of such an approach is even more evident in view of a mathematical fable which is dual, in some vague sense, to the Achilles and the Tortoise paradox (but perhaps this duality could be made explicit). It is told in Harvey M. Friedman's lectures *Philosophical Problems in Logic* [35]. Friedman said:

> I have seen some ultrafinitists go so far as to challenge the existence of 2^{100} as a natural number, in the sense of there being a series of "points" of that length. There is the obvious "draw the line" objection, asking where in
>
> $$2^1, 2^2, 2^3, \ldots, 2^{100}$$

we stop having "Platonistic reality". Here this ... is totally innocent, in that it can be easily replaced by 100 items (names) separated by commas. I raised just this objection with the (extreme) ultrafinitist Yessenin-Volpin during a lecture of his. He asked me to be more specific. I then proceeded to start with 2^1 and asked him whether this is "real" or something to that effect. He virtually immediately said yes. Then I asked about 2^2, and he again said yes, but with a perceptible delay. Then 2^3, and yes, but with more delay. This continued for a couple of more times, till it was obvious how he was handling this objection. Sure, he was prepared to always answer yes, but he was going to take 2^{100} times as long to answer yes to 2^{100} than he would to answering 2^1. There is no way that I could get very far with this.

Yessenin-Volpin's response makes it clear that the Achilles and the Tortoise paradox is not so much about actual infinity as of a potential infinity (or just plain technical feasibility) of producing the sequence

$$\frac{1}{2}, \frac{1}{4}, \frac{1}{8}, \frac{1}{16}, \text{ etc.,}$$

in real time.[6]

However, there is yet another layer to this story. It provides an opportunity to bring into the discussion a rarely mentioned aspect of mathematical practice: the influence of the personality of a mathematician on his or her mathematical outlook.

The instantaneousness of Yessenin-Volpin's response to the line of questioning is more than a quick reflex. One should remember that Alexander Yessenin-Volpin was one of the founding fathers of the Soviet human rights movement and spent many years in prisons and exile. He knows a thing or two about interrogations; in 1968, he wrote and circulated via Samizdat the famous *Memo for those who expect to be interrogated*, much used by fellow dissidents.

One piece of advice from the *Memo* is worth quoting:

> During an interrogation, it is already too late to determine your position and develop a line of behavior. [...] If you expect an interrogation, if there is just a possibility of an interrogation—get prepared in advance.

Yessenin-Volpin was also a poet of note. One of his poems, a very clever and bitterly ironic rendition of Edgar Allan Poe's *The Raven*, is quite revealing in the context of our discussion. I give here only the first two and the last three lines of the poem.

Как-то ночью, в час террора, я читал впервые Мора,
Чтоб Утопии незнанье мне не ставили в укор ...

$$\vdots$$

Но зато как просто гаркнул черный ворон: "Nevermore!"
И качу, качу я тачку, повторяя: "Nevermore..."
Не подняться ... "Nevermore!"

To make these lines more friendly to the English speaking reader, I explain that the first two lines refer to Thomas More's *Utopia*: the protagonist reads *Utopia* to avoid an accusation that he has not familiarized himself with the utopian teachings promoted by the totalitarian system. The three exclamations "Nevermore!" which end the poem do not need translation.

The poem was written in 1948 (significantly, the year when George Orwell wrote his *1984*—the title of the novel is just a permutation of digits; in 1949, when Orwell's novel was published, Yessenin-Volpin started his first spell in prisons). As we can see, Yessenin-Volpin, who was 23 years old at the time, developed an ultrafinitist approach to utopian theories (and even more so to utopian practices) much earlier than to problems of mathematical logic.

5.5 The vanishing point

> Father Ted: *Now concentrate this time, Dougal. These*
> (he points to some plastic cows on the table)
> *are very small; those*
> (pointing at some cows out of the window)
> *are far away ... Small. Far away.*
> Father Ted [443]

In the discussion of the psychology of mathematics, the problem of infinity is unavoidable; our visual perception of infinity is particularly puzzling; I mentioned it in Section 5.1; here, I wish to return to the subject.

I stated earlier in this chapter that our first encounter with infinity does not happen when we see railway tracks merging at the horizon; the reason for this is that our brains resolutely refuse to recognize that the parallel rails converge at the vanishing point. Instead, your brain reminds you, like Father Ted to Father Dougal in the famous episode of the cult TV series, that, although the cow in the field projects onto a tiny spot on the retina in your eye, while a toy cow in your hands projects onto a much larger retina image, it is the cow in the field that is actually bigger. Similarly, your brain continues to remind you that the tracks are actually parallel all the way to the horizon. The brain actively fights geometry!

The ability to use and read linear perspective in simple schematic drawings is the result of cultural conditioning. (It is different with photographs where the perspective, with its vanishing points, is present due to the *physical* laws of geometric optics.) In ancient Egypt, for example, the graphic culture was quite different:

> In seeking to represent three-dimensional objects on a plane surface, whether a drawing board or an area of the wall, the Egyptian avoided the perspectival solution of the problem which alone of the nations of antiquity, the Greeks ultimately reached by the fifth century B.C. Their vision of the world, seen from a certain standpoint at a certain moment of time, would have seemed to the ancient Egyptian as presumptuous, and concerned only with illusion, a mere distortion of reality. The Egyptian was concerned not with presenting an evanescent personal impression, caught in an instant, but with what he regarded as eternal verities. [...]
>
> His non-perspectival vision placed the Egyptian artist in harmony with the world that he knew to exist. His perception of the forms of nature was derived from a fusion of several aspects recollected in the tranquility of his mind and not captured as an instant revelation to the seeing eye. [422, pp. 13–15]

In the Eastern Orthodox tradition of icon painting (which originated in Byzantium), for example, an over-elaborate and technically sophisticated system of *reverse perspective* is employed: the parallel lines intersect in the foreground of the painting, creating a very soothing, comforting feeling of a finite world which embraces the viewer.

This paradox is even more remarkable because the image on your retina, in accordance with the simple laws of geometric optics, is similar to photographic images and is something like the one in Figure 5.6, with tracks meeting at the vanishing point on the horizon. But, in the environment which shaped the evolution of our ancestors, the vanishing point was of no importance; for survival, it was much more important to recognize that a tiny speck of grey on your retina was the Big Bad Wolf.

One way to make your brain accept the existence of the vanishing point is to dramatically simplify its task of visual processing. The best feeling of geometric perspective is achieved when you look into a long dark tunnel with a tiny spot of daylight at the opposite end or (as I did when I was a child) stare into a deep, deep well, with a tiny reflection of blue sky in the water. The brain is no longer forced to recalculate the actual sizes of objects, because there are not any in the field of view; the vanishing point, which was of no importance, becomes the only source of light and dominates the field of view.

Fig. 5.6. The railway tracks converging at the vanishing point. ©2003 by Tomasz Sienicki, licensed under Creative Commons Attribution 2.5 License. Source: *Wikipedia Commons*.

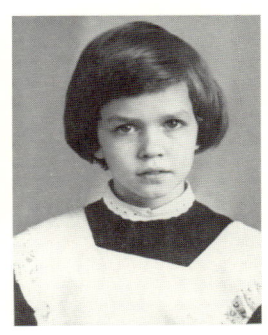

Inna Korchagina, aged 8

The same happens in the "tunnel of light" illusion (or hallucination) caused by the shutting down of light receptors in the retina, as a result of oxygen deprivation or the effects of drugs. I experienced it as a boy when I broke my arm in a skiing accident and was given rather barbaric ether narcosis; the last thing I remembered about the real world was the surgeon saying to the nurse: "Add a bit more"—and ether dripping from the gauze mask on my face. Then the flight started. In a rational reconstruction, the light sensors shut down one by one from the periphery of the retina to its center; this is perceived as if only the receptors at the center of the retina receive light and is deciphered by the brain as the light at the end of the tunnel. Adding to that illusion, the vestibular apparatus, knocked out by ether, reports to your brain the sensation of free fall, while the feeling of time is also suppressed—and you have the astonishing, overwhelming out-of-body experience of a flight through the endless tunnel towards the Light—and into Nothingness.

Paradoxically, in order to recognize infinity in the form of the vanishing point of perspective, your brain has to be so seriously impaired that infinity is no longer mathematical.

Not surprisingly, the "tunnel of light" could be a powerful religious and spiritual experience and, as such, is well documented—see *Paradisio*, Gustave Doré's illustration to Dante's *Divine Comedy*, Figure 5.7. But religious and spiritual experiences are strictly individual and not reproducible—and this draws the line which

Fig. 5.7. *Paradisio* (plate 35) by Gustave Doré—Illustration to Paradiso by Dante Alighieri, Canto 31, Verse 1–3. Source: *Wikipedia Commons*. Public domain.

separates infinity, and how it is understood and perceived both formally and informally in mathematics, from its religious and spiritual interpretation.

5.6 How humans manage to lose to insects in mind games

To further discuss the surprising inadequacy of the human perception of perspective, I turn to a fascinating study of *motion camouflage*, based on papers by my colleague Paul Glendinning [183, 184].

Dragonflies, elegant creatures much beloved by poets and children, are consummate predators. As so frequently happens with

Fig. 5.8. Dragonfly. Source: Wikipedia Commons. Public domain.

beautiful predators, their real sophistication is reached not in delightful air acrobatics but in the ways they hunt and fight.

In their fights for territory, dragonflies, when they pursue the enemy with the aim of a sudden attack, use a remarkable and inventive concealment strategy. A dragonfly camouflages its approach so that the foe believes it to be stationary. This conclusion is the result of the reconstruction of stereo camera images carried out by Mizutani, Chahl, and Srinivasen [207]. There are already several studies suggesting possible guidance mechanisms used by dragonflies [149, 232]; however, the underlying mathematical problem is not that sophisticated. Indeed, I quote Paul Glendinning:

> This is not as hard as it sounds. Even when moving, most animals have a good sense of the direction to a given fixed object at any time, and expect to see it on that line. If the aggressor moves so that at each moment it is on the line between the target and a given fixed point, which could be its initial position, then its relative motion in the eyes of the target is the same as that of the stationary reference point. The only way that the target can know that it is not stationary is to notice the change in size of the aggressor as it approaches. Mizutani, Chahl and Srinivasen [207] extrapolate the lines between the aggressor and the target at several different times and show that to a good approximation these all meet at a point, the fixed reference point or initial condition of the aggressor.

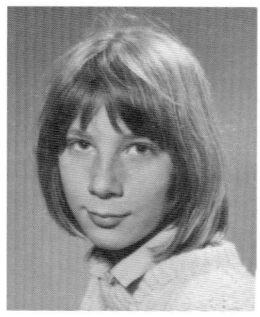

Paul Glendinning,
aged 12

In a development highly offensive to humans' pride, it seems that we too can be duped by dragonflies and hover-flies: this is confirmed by experimental studies [150] and resolves the mystery of dragonfly flight which puzzled me when I was a child. When you see a dragonfly over a river or meadow, it frequently appears to be hanging in the air motionless, as if it were glued to the sky, and then it suddenly jumps at you, whizzing by a few inches from your face—you feel the air stream from its wings on your skin. Only upon reading Glendinning's paper did I realize that the dragonfly was not motionless—it was approaching me on a reconnaissance flight.

If you consider for a second how such an astonishing defeat in a mathematical game against *insects* could ever happen, it becomes apparent that, although human vision is exceptionally good at detecting even the tiniest relative changes in the position of an object, we are not good at detecting gradual increases in the relative *size* of an object. (David Broomhead suggests that this can be explained by the structure of light receptors in our eyes' retina.) You would probably agree with me if you have ever looked at a distant approaching train, seen as a spot of light at the vanishing point of the rail tracks. It is very hard to judge whether the train is stationary or if it is approaching the platform. It is even harder in modern Britain, where it has become a non-trivial proposition.

Humans use similar strategies, sometimes developed individually, using trial and error at a semi-conscious level, or perhaps learned as part of professional training. For example, baseball players apparently catch high balls by running along such paths on the field—and with varying speed—that the ball (which they keep continuously in view) is perceived as hanging motionless in the sky.

Another example comes from seafaring practice, as a criterion used in sailing to detect boats on a collision course.[7] I again quote Glendinning:

> If a boat appears to be stationary with respect to some distant reference point or has the same compass bearing from your boat over a period of time, then it is on a collision course with you. [441]

This is equivalent to active motion camouflage with the reference point at infinity.

Notice: geometric infinity has again appeared on the scene. Of course, from a geometric perspective, the reference point at infinity is the same as the vanishing point. Unlike the verbal infinity of counting, it requires much longer and more arduous training.

5.7 The nightmare of infinitely many (or just many) dimensions

> *The miserable wasteland of multidimensional space was first brought home to me in one gruesome solo lunch hour in one of MIT's sandwich shops. "Wholewheat, rye, multigrain, sourdough or bagel? Toasted, one side or two? Both halves toasted, one side or two? Butter, polyunsaturated margarine, cream cheese or hoummus? Pastrami, salami, lox, honey cured ham or Canadian bacon? Aragula, iceberg, romaine, cress or alfalfa? Swiss, American, cheddar, mozzarella, or blue? Tomato, gherkin, cucumber, onion? Wholegrain, French, English or American mustard? Ketchup, piccalilli, tabasco, soy sauce? Here or to go?"*
>
> Myles Aston [439]

Very frequently, when we deal with a mathematical object and wish to modify it and make it "infinite" in some sense, we have several different ways for doing so. For example, usual decimal numbers can be extended to infinite decimal expansions to the *right*:

$$\pi = 3.1415926\ldots$$

and to the *left*:

$$\ldots 987654321$$

In the second case, the operations of multiplication and addition are defined in the usual way, with the excess carried to the next position on the left (which, by the way, is more natural than the multiplication of infinite decimal fractions[8]). With these operations we get the so-called ring of 10-adic integers. Properties of 10-adic integers are quite different from that of real numbers: to give one example, you cannot order 10-adic integers in a way compatible with addition and multiplication (so that the usual rules of manipulating inequalities would hold). This can be seen already from one of the simplest instances of addition:

$$\ldots 99999 + 1 = \ldots 00000 = 0.$$

10-adic integers are not frequently used in mathematics, but p-adic integers for prime values of the base p, defined in a similar way by expanding integers written to base p to the left, are quite useful and popular.[9] I notice, in passing, that a paradoxical summation of the infinite series

$$1 + 2 + 4 + \cdots = -1,$$

due to Euler, makes sense and is completely correct in the domain of 2-adic integers, written by base 2 expansions. Indeed, it becomes an easy-to-check arithmetic calculation:

$$\ldots 11111 + 1 = \ldots 00000 = 0.$$

Well, this example is still "tame" infinity, encapsulation of the process indexed by the ordered set of natural numbers, and is, in some vague sense, one-dimensional.

The really hard-to-comprehend infinity arises in situations when we try to increase indefinitely the number of dimensions in the problem. There are some pretty obvious reasons why this could be difficult.

- Our intuition about the properties of individual objects starts to fail us very soon; we lose the mental picture of the "general" term.
- On top of that, the change of dimension is frequently too dramatic a change in the object; we easily lose the mental picture of the "general step" in the process.
- The resulting mathematical object frequently has properties dramatically different from that of any of its finite-dimensional analogues.

The following example is taken from the book *Mathematical Aquarium* by Victor Ufnarovski [292].

Inscribe four equal circles of maximal possible size into a square with side 1, and a small circle in the center touching all four:

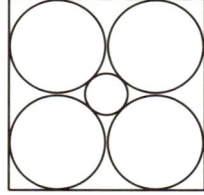

Similarly, you can consider 8 spheres inscribed into the cube of side 1, with the small sphere in the center touching all 8, then $2^4 = 16$ small spheres inscribed into the four-dimensional cube, etc. What is the limit of the radii r_n of central spheres touching 2^n equal spheres inscribed into the n-cube of side 1 as n tends to infinity?

The answer is paradoxical: infinity. The explanation is very simple: the main diagonal of the n-cube (the one that connects the opposite vertices) has length $\sqrt{n}$ by the repeated use of the Pythagorean Theorem and grows to infinity, which is already disturbing on its own. Consider how the central sphere of unknown radius r_n and two spheres of (unchanging) radius $1/4$ sit on the very long diagonal for large n:

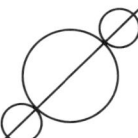

I hope that the conclusion is now obvious. In particular, this means that, at some n, the radius of the central sphere becomes bigger than $1/2$ and the sphere starts to stick out of the cube! I leave it to the reader as an exercise to find the value of the dimension n when this first happens. And one more question to train your multidimensional intuition: does the central sphere ever intersect the (one-dimensional) edges of the cube?[10]

This toy example is an elementary expression of a well-known basic fact of analysis. In n-dimensional Euclidean space, you can introduce a topology and concept of limits in two different ways which happen to be equivalent. In one case, you use small spheres to measure the closeness of points to each other; this is equivalent to using the usual formula of Euclidean geometry

$$\sqrt{x_1^2 + \cdots + x_n^2}$$

to measure the length of a vector with coordinates $(x_1, \ldots, x_n)$. In the other approach, you take the so-called *uniform norm*

$$\|\vec{x}\| = \max\{|x_1|, \ldots, |x_n|\},$$

Gábor Megyesi, aged 15

so the "sphere"

$$\{\, \vec{x} \mid \|\vec{x}\| \leqslant 1 \,\}$$

is actually a cube, and you use small cubes to measure the closeness of points to each other. But, as I have already said, the topologies are the same: inside of every sphere one can place a smaller cube, and vice versa.

In the commonly used infinite-dimensional limit, vectors are infinite sequences

$$x_1, x_2, x_3, \ldots$$

such that the sum under the radical in the expression for Euclidean distance converges,

$$\sum_{i=1}^{\infty} x_i^2 < \infty;$$

then its square root,

$$\sqrt{\sum_{i=1}^{\infty} x_i^2},$$

can be taken for the "length" of the vector. On the other hand, we can take for the "length" of the vector its *uniform convergence* norm

$$\|\vec{x}\| = \limsup\{|x_1|, |x_2|, |x_3|, \ldots\}.$$

Unlike the case of finite-dimensional spaces, the corresponding "sphere" and "cube" topologies (known in mathematics as the l^2 topology and the uniform convergence topology) and the corresponding theories of limits are quite different.[11]

Notes

[1]Escher has explicitly used regular tessellations of the hyperbolic plane as a framework for his hypnotic patterns of interlocking motifs—a point made obvious by Douglas Dunham [28] in his experiments with Escher's patterns: it is possible to take Escher's motif and render it into a different tessellation; see Figure 5.4.

[2]DIVISION BY THREE. To scare the reader into acceptance of the intrinsic difficulty of division, I refer to the paper *Division by three* [345] by Peter Doyle and John Conway. I quote their abstract:

> We prove without appeal to the Axiom of Choice that for any sets A and B, if there is a one-to-one correspondence between $3 \times A$ and $3 \times B$, then there is a one-to-one correspondence between A and B. The first such proof, due to Lindenbaum, was announced by Lindenbaum and Tarski in 1926, and subsequently 'lost'; Tarski published an alternative proof in 1949. We argue that the proof presented here follows Lindenbaum's original.

Here, of course, 3 is a set of 3 elements, say, $\{0, 1, 2\}$.

Prove this in a naive set theory *with* the Axiom of Choice.

[3]EIDETIC IMAGINATION. My friend Owl told me that, in her adult life, she continued to experience, occasionally, mental images of eidetic intensity, mostly related to music or mathematics. For example, during her first year at the university she had an eidetic experience concerned with the Dirichlet function. The Dirichlet function, by definition, takes value 1 at

rational points and 0 at irrational points—it is not something which is easy to visualize.

[4]COMPLETED ACTIONS. One of the earlier readers and critics of this book, Gregory Cherlin, commented that some human languages have grammatical markers for *completed actions* as opposed to *uncompleted*, which perhaps might make it easier for a learner to encapsulate the actual infinity—not unlike how plurality markers help a learner to grasp the concept of number (Section 3.2).

[5]ARCHIMEDES' AXIOM. The property known as Archimedes' Axiom can be traced to the Preface of Archimedes' *Quadrature of the Parabola* where he says that any area

> "can, if it be continually added to itself, be made to exceed any assigned finite area".

[6]ACCELERATED TURING MACHINES. Time considerations of a Zeno Paradox type are indeed closely intertwined with abstract concepts of computation. Let us consider an idealized sci-fi Turing machine which makes every iteration twice as quick as the previous one. Then even the hardest functions become computable in subexponential time, and uncomputable functions become computable. This idea is developed in considerable technical detail by Jack Copeland in his paper [332]. Copeland attributes the idea of 'acceleration' to Bertrand Russell:

> Miss Ambrose says it is logically impossible [for a man] to run through the whole expansion of π. I should have said it was medically impossible. ... The opinion that the phrase 'after an infinite number of operations' is self-contradictory seems scarcely correct. Might not a man's skill increase so fast that he performed each operation in half the time required for its predecessor? In that case, the whole infinite series would take only twice as long as the first operation. [82, pp. 143–144]

He also attributes it to Hermann Weyl who proposed considering an imaginary machine capable of carrying out

> an infinite sequence of distinct acts of decision within a finite time, say, by supplying the first result after $1/2$ minute, the second after another $1/4$ minute, the third $1/8$ minute later than the second, etc. In this way it would be possible ... to achieve a traversal of all natural numbers and thereby a sure yes-or-no decision regarding any existential question about natural numbers. ([98, p. 34] or p. 42 of the English translation)

See also [333, 334].

[7]COLLISIONS. Here is a classical problem on collisions from *Littlewood's Miscellany* [65]:

4 ships A, B, C, D are sailing in fog with constant and different speeds and constant and different courses. The five pairs A and B, B and C, C and A, B and D, C and D have each had near collisions; call them 'collisions'. Most people find unexpected the mathematical consequence that A and D necessarily 'collide'.

Prove that! The problem can be classified as belonging to projective geometry, a mathematical discipline (and the class of mathematical structures) with historic origins in the study of geometric perspective.

[8]THE TROUBLE WITH REAL NUMBERS. If you still believe that real numbers are the best of all worlds, try to find, without a calculator, the first three digits of the ratio

$$\frac{0.12345\cdots484950}{0.5152\cdots99100}.$$

In principle, the first few digits of the numerator and the denominator should suffice for the computation. But how many of them do you need?

[9]n-ADIC INTEGERS AND INTEGER DIVISORS. 10-adic integers are not as good as p-adic integers for prime p because they contain zero divisors, non-zero numbers x and y such that $xy = 0$. The following elementary example was provided by Hovik Khudaverdyan and Gábor Megyesi. If you look at the sequence of iterated squares

$$5,\ 5^2 = 25,\ 25^2 = 625,\ 625^2 = 390625,\ 390625^2 = 152587890625\ldots,$$

you notice that consecutive numbers have an increasingly long sequences of the rightmost digits in common, that is,

$$5^{2^{n+1}} \equiv 5^{2^n} \mod 10^n,$$

a fact which could be easily proven by induction. Hence the sequence converges to a 10-adic integer

$$x = \ldots 92256259918212890625$$

which has the property that $x^2 = x$ and hence $x(x - 1) = 0$.

One can see that zero divisors appear in the ring of 10-adic integers because 10 is not a prime number. An exercise for the reader: prove that the ring of 2-adic integers has no zero divisors.

[10]THE UNIT CUBE. Zong [420, 421] gives a comprehensive survey of known and still conjectural properties of the n-dimensional unit cube.

[11]MORE ON THE GEOMETRY OF THE UNIT CUBE. I conclude my story with yet another fable, an example of (incomplete) induction failing because of the paradoxical geometry of the n-cube. See Borwein [322] for the explanation of the sudden jump in behavior of a series of integral identities—it is essentially the same effect as in Ufnarovski's inscribed spheres problem.

Define

$$\operatorname{sinc}(x) = \sin(x)/x, \qquad x \neq 0,$$

and

$$\operatorname{sinc}(0) = 1.$$

One can prove the following identities (details are in [322]):

$$\int_0^\infty \mathrm{sinc}(x) = \pi/2,$$

$$\int_0^\infty \mathrm{sinc}(x)\mathrm{sinc}(x/3) = \pi/2,$$

$$\int_0^\infty \mathrm{sinc}(x)\mathrm{sinc}(x/3)\mathrm{sinc}(x/5) = \pi/2,$$

$$\int_0^\infty \mathrm{sinc}(x)\mathrm{sinc}(x/3)\mathrm{sinc}(x/5)\mathrm{sinc}(x/7) = \pi/2,$$

$$\int_0^\infty \mathrm{sinc}(x)\mathrm{sinc}(x/3)\mathrm{sinc}(x/5)\mathrm{sinc}(x/7)\mathrm{sinc}(x/9) = \pi/2.$$

It appears that a certain pattern emerges. But let us continue:

$$\int_0^\infty \mathrm{sinc}(x)\mathrm{sinc}(x/3)\mathrm{sinc}(x/5)\mathrm{sinc}(x/7)\mathrm{sinc}(x/9)\mathrm{sinc}(x/11) = \pi/2,$$

$$\int_0^\infty \mathrm{sinc}(x)\mathrm{sinc}(x/3)\mathrm{sinc}(x/5)\mathrm{sinc}(x/7)\mathrm{sinc}(x/9)\mathrm{sinc}(x/11)\mathrm{sinc}(x/13) = \pi/2.$$

Then, out of blue:

$$\int_0^\infty \mathrm{sinc}(x)\mathrm{sinc}(x/3)\mathrm{sinc}(x/5)\mathrm{sinc}(x/7)\mathrm{sinc}(x/9)\mathrm{sinc}(x/11)\mathrm{sinc}(x/13)\mathrm{sinc}(x/15)$$

$$= \frac{467807924713440738696537864469}{935615849440640907310521750000} \cdot \pi.$$

It is dangerous to hold a formula true after checking just a few first instances, especially if the formula changes its shape, not just the parameter values!

6

Encapsulation of Actual Infinity

6.1 Reification and encapsulation

Затоварилась бочкотара, затарилась,
затюрилась и с места стронулась.
Vasilii Aksenov

Further development of the principal themes of this book is impossible without moving closer to the established methodological framework of mathematics education theory. Therefore, I turn to the discussion of *reification*. There is a significant body of literature, both theoretical and experimental studies, which deals with reification mostly in the framework of school mathematics teaching.

The term *reification* was introduced into mathematics education studies by Anna Sfard, who applied it to the process of objectivization of mathematical activities. The concept is pretty close to that of *encapsulation* [142]. One may wish to find subtle differences in the meanings of the two concepts. But, since in application to real case studies they become blurred anyway, I refrain from taking the possible differences into account.

The associated verb is *to reify*, with the meaning "to convert mentally into a thing, to materialize". In Marxist literature, the term "reification", as well as its more specialized version, *commoditization*, has rather negative connotations, which are absent in Sfard's use of the word.

The concept of reification is exceptionally useful in the understanding of mathematics teaching and learning. According to Reuben Hersh's succinct description, children first learn an activity, something they do; this activity is frequently formalized as an algorithm but sometimes remains semi-formal. Later the activity becomes a "thing", something they can think about as an object. This "reification" step is difficult for a student (see its discussion in the dialog between Anna Sfard and

Anna Sfard,
aged 7

Pat Thompson [130]) and is a main contributing factor to the success or failure of mathematics teaching.

The term "encapsulation" is sometimes more convenient because it allows us to define a natural opposite action which we shall call *de-encapsulation*; "de-reification" sounds odd. Also, the term "encapsulation" is better suited for situations when the process is intentional and deliberate, as in a work of a research mathematician. But I would prefer to reserve "reification" for the description of an amorphous, undirected (but perhaps guided by a teacher) process of emergence of a concept in mathematics learning, especially at the earlier stages of mathematical education.

This is the description of encapsulation and de-encapsulation in Weller et al. [142, p. 744]:

> The encapsulation and de-encapsulation of a process in order to perform actions is a common experience in mathematical thinking. For example, one might wish to add two functions f and g to obtain a new function $f + g$. Thinking about doing this requires that the two original functions and the resulting function are conceived as objects. The transformation is imagined by de-encapsulating back to the two underlying processes and coordinating them by thinking about all of the elements x of the domain and all of the individual transformations $f(x)$ and $g(x)$ at one time so as to obtain, by adding, the new process, which consists of transforming each x to $f(x) + g(x)$. This new process is then encapsulated to obtain the new function $f + g$.

It is instructive to see how Anna Sfard assesses a mathematician's description of his work. In [130] she quotes a famous mathematician, Bill Thurston:

> Mathematics is amazingly compressible: you may struggle a long time, step by step, to work through some process or idea from several approaches. But once you really understand it and have the mental perspective to see it as a whole, there is a tremendous mental compression. You can file it away, recall it quickly and completely when you need it, and use it as just one step in some other mental process. The insight that goes with this compression is one of the real joys of mathematics. [139, p. 847]

Sfard comments on this:

> If the "compression" is construed as an act of reification—as a transition from operational (process-oriented) to structural vision of a concept . . . , this short passage brings in full relief the most important aspects of such transition. First,

it confirms the developmental precedence of the operational conception over the structural: we get acquainted with the mathematical process first, and we arrive at a structural conception only later. Second, it shows how much good reification does to your understanding of concepts and to your ability to deal with them; or, to put it differently, it shows the sudden insight which comes with "putting the helmet and glove on" [Sfard uses here her "virtual reality game" metaphor; see Section 12.2.–AB] with the ability to see objects that are manipulated in addition to the movements that are performed. Third, it shows that reification often arrives only after a long struggle. And struggle it is!

I would not construe Thurston's words in the same way. What he describes is *not* reification. More precisely, reification is present in the process but is only a tiny portion of it. Even where reification is present, it is directed by mathematical structures, by metatheories, and is quite purposeful and intentional. The process of compression, as described by Thurston, also involves a systematic search for new languages or translation of the problem into other known languages, meta-arguments and analysis of existing proofs, etc. To call all these actions reification is to stretch the useful concept to the point where it becomes all-embracing and vacuous. Moreover, reification itself is frequently compressed, in the same way as other mathematical activities tend to compress themselves into *reusable* (and sometimes even *reproducible*) units.

6.2 From potential to actual infinity

> *It does not matter if a cat is black or white,*
> *as long as it catches mice.*
> Deng Xiao-ping

We discussed potential infinity in Section 5.3. Actual infinity is harder; to accept the actual infinity of the set of natural numbers, as the *final result of counting by ones*, to think of *all* natural numbers *together* as one infinite set, is a leap of faith. Not everyone, even a professional mathematician, is prepared to make it or will admit to making it; the list of sceptics includes great names, such as Henri Poincaré. But, once established, it is passed from generation to generation of mathematicians like a religious belief; general cultural influences can make certain mathematical concepts appear self-evident.[1] Indeed, it is much easier to create an eidetic image of something when you are told that this something exists and are given a name for it. It is easy to imagine a unicorn after someone has described it to you in detail. We have to admit that some degree of coercing and cheating is normal in teaching mathematics: there is something of the unicorn in many mathematical concepts as we teach them to our students.

I see nothing wrong with that; to conceive the concept of actual infinity is a great discovery (or revelation); as teachers, we would be naive if we expected every one of our students to repeat one of the most dramatic feats of human intellect. We simply have to offer actual infinity to them in ready-to-use form—and convince them, at both the emotional and practical levels, that it is a useful concept, that it brings reproducible results.

As a child, I was told to learn by heart the times tables, without any attempts to explain to me where they came from. I can only regret that my teacher was not Vladimir Radzivilovsky (see Section 4.6) who encouraged every child to *compose* the times tables by adding numbers by twos, threes, etc., and only then memorize them. However, whatever the origin of the times tables was, I have been using them all my life—and, you know, they work!

Sergei Konyagin,
aged 15

But there are better ways to guide students than intimidation and abuse of the teacher's authority.

6.2.1 Balls, bins, and the Axiom of Extensionality

> *Man is equally incapable of seeing*
> *the nothingness from which he emerges*
> *and the infinity in which he is engulfed.*
> Blaise Pascal

My case study of interiorization of actual infinity was suggested by a recent paper by Weller et al. [142].[2] The authors analyze students' approaches to the resolution of a classical paradox:

> Suppose you put two tennis balls numbered 1 and 2 in Bin A and then move ball 1 to Bin B, then put balls 3 and 4 in Bin A and move 2 to Bin B, then put balls 5 and 6 into Bin A and move 3 to Bin B, and so on without end. How many balls will be in Bin A when you are done?

Instead of trying to resolve it on the spot—either as a mathematical or pedagogical problem—I suggest making three new problems (with, possibly, quite different solutions) from it. Please notice that I do not care about solutions for the new problems; their only purpose is to shed some light on the old one.

The first problem deals with *indistinguishable* balls:

> Suppose you put two tennis balls in Bin A and then move, at random, one ball to Bin B, then put two new balls in Bin

A and move a random ball from Bin A to Bin B, and so on without end. Will Bin A be empty when you are done? [?]

> *Still, this is a good problem: try to solve it.*

I love this form of the paradox because it can be shown that every individual ball ends up in Bin B with probability 1. [?] But *Why?* does that mean that Bin A is empty?

The second formulation is a continuous version of the indistinguishable balls problem (we first think of balls as molecules of water and then, as physicists do, ignore the molecular structure of water and think of it as continuous matter).

> Suppose you have two tanks A and B and you pour water into tank A at constant rate; meanwhile, water leaks from A to B at smaller constant rate, and so on without end. How much water is in tank A when you are done?

The third problem replaces the balls with quantum particles, say electrons. Then not only are individual electrons indistinguishable, but their locations cannot be specified, so that we can talk only about the expected numbers of particles in Bin A and Bin B.

If these three problems still do not provide enough food for thought, we may recall that we have a number of natural questions which need to be addressed before we present the paradox to the students. For example, the Bin Problem, in its original formulation, amounts to computing the value of the expression

> *It frequently happens in mathematics, that, in order to make a concept intuitive, we have to jump to the next level of abstraction.*

$$\{1,2\} \smallsetminus \{1\} \cup \{3,4\} \smallsetminus \{2\} \cup \cdots .$$

Have we not told the students that infinite expressions, like the sum

$$2 - 1 + 2 - 1 + 2 - 1 + \cdots ,$$

are meaningless unless what they mean is explicitly defined?[3]

As frequently happens in mathematics, in order to make a concept intuitive, we have to jump to the next level of abstraction; you cannot expect that from students, but a teacher can and should do that for them. Instead of trying to help our students to encapsulate *one* iterative process which produces a potentially infinite stream of elements into the set of *all* elements in the stream, we can start confidently talking about *arbitrary sets*, ignoring as irrelevant the process which produced them.

The analysis of the examples above makes it abundantly clear that what really matters is that sets are composed of distinguishable elements, with every element having an identity of its own.

To understand sets as *objects*, we need some tools for manipulating them; an object is not an object if we just look at it and do nothing. To *compare sets*, we need to postulate the *Axiom of Extensionality* (sometimes called the *Volume Principle*):

> Two sets A and B are equal if and only if each element of either set is an element of the other.[4]

Notice that the word "infinity" is never mentioned in the definition of equality of sets; the actual process (quite possibly infinite) of checking, element by element, that they belong to B is also not mentioned. The Volume Principle is a great example of *abstraction by irrelevance* as discussed in Section 7.8.

This is an exceptionally deep principle: infinity creeps into it through the back door. For example, (potentially) infinitely many possible definitions of the empty set lead to the same set, THE unique empty set! As Brian Butterworth [159] nicely put it:

> Although the idea that we have no bananas is unlikely to be a new one, or one that is hard to grasp, the idea that no bananas, no sheep, no children, no prospects are really all the same, in that they have the same numerosity, is a very abstract one.

To this I add that "no bananas" and "no prospects" do not just have the same numerosity, they ARE the same; the equality of their numerosities is a mere corollary.

After that, the question

> Is the final set of tennis balls in Bin A empty?

has a simple answer: YES!

Being brought up in a Hegelian philosophical tradition, I see paradoxes of actual infinity as manifestations of the Hegelian dialectical transition; as all dialectical contradictions, the chasm between finite and infinite is relative and could—and should—be *sublated*, removed by change of a viewpoint, the same way as we remove the mold from the cast or scaffoldings from the finished building.

As a teacher, I see my task not as encouraging students to bang their heads against the wall but as providing alternative viewpoints which eliminate the paradox.

Finally, I wish to mention a remark by Gregory Cherlin: it is likely that comprehension of actual infinity is easier for people whose native tongues makes a clear distinction between verbs for complete and incomplete action.

6.2.2 Following Cantor's footsteps

One of the possible ways to give students a good intuitive feeling of actual infinity is via a bit more detailed discussion of equations and their solutions: for example, the solution set of

$$\sin\left(e^{\tan x}\right) = 0 \tag{6.1}$$

for positive real x is well ordered (that is, every subset contains a smallest element) and, in terminology of set theory, has ordinal type ω^2. This expression means that it looks like the set of natural numbers (which is said to have ordinal type ω) repeated infinitely many times, more precisely, ω times, once for every natural number.[5] When solving (6.1), the standard process of listing the roots of $\sin x = 0$,

$$0, \pi, 2\pi, 3\pi, \ldots,$$

has to be repeated infinitely many times, for each branch of $\tan x$ (see Figure 6.1). We have a process made of processes; to handle it, most students (well, those who can handle the equation in the first instance) will have no choice but to start thinking of individual smaller processes as *objects*, that is, to encapsulate them.

This example to some degree reproduces Georg Cantor's first steps in his creation of set theory: he was motivated by his work on convergence of trigonometric series where he had to somehow deal with ensembles or collections of points where the series diverges and, in particular, introduced the concept of an ordinal number or type. Ordinals preceded cardinals!

6.2.3 The art of encapsulation

The example with encapsulation of actual infinity suggests some basic principles of encapsulation:

- If possible, never encapsulate on the basis of a stand-alone example. Always work in a wider set-up, where individual processes (or raw pre-encapsulation concepts) interact with each other and give you no choice but to start thinking of them as objects.
- Be attentive to the details of interaction: they will help you to find the right mathematical formulation of the encapsulated concept.
- And my last piece of advice is taken from the *999* TV series (BBC accident and rescue reconstructions):
    if you suffer an attack of vertigo—do not freeze: move.

We will return to our discussion of encapsulation in Section 6.3.

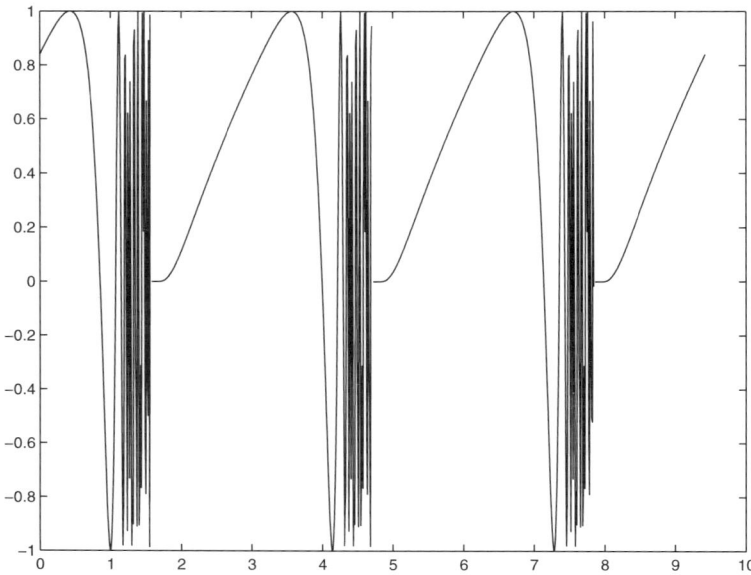

Fig. 6.1. The graph of $y = \sin\left(e^{\tan x}\right)$ as produced by MATLAB by plotting 300 points. The inevitably approximate nature of the plot causes numerous deviations from the actual shape of the curve: all peaks should have value 1, of course.

6.2.4 Can one live without actual infinity?

How should we approach the position of those mathematicians who refused to accept actual infinity? Henri Poincaré is a prominent example. Weller et al. [142] quote him on a par with modern day undergraduate students:

> "There is no actual infinity; and when we speak of an infinite collection, we understand a collection to which we can add new elements unceasingly."

If I had in my class a student who made a similar statement and if the student's name happened to be Henri Poincaré, I would not be in a hurry to correct him but would try to check whether Henri had a consistent vision of mathematics which was compatible with his thesis. For example, in his understanding, did an algebraic curve given by equation $f(x, y) = 0$ *consist* of points or was it a *locus*, placeholder for points?

In modern terminology, the first point of view is set-theoretic, the second one is scheme-theoretic (and is based on the concept of a category, not a set). Modern geometry works mostly with categories, not sets, and the archaic, pre-set-theoretic concepts of geometric objects fit very smoothly into the modern framework.

Also, I have a gut feeling that Henri would still be able to solve every problem in the advanced calculus/analysis class (including the one from Section 6.2.2). Indeed, within undergraduate calculus, every problem is solvable within the framework of the potential infinity of processes and sequences—you need only to be sufficiently attentive to detail. This is not as difficult as it seems—in order to deal with sequences and series, for example, you have, anyway, to de-encapsulate them back to their finite initial sequences and partial sums.

6.2.5 Finite differences and asymptotic at zero

I hope I will be excused for turning to my own, highly unorthodox, learning experience. It so happened that I was a guinea pig in a bold educational experiment: at my boarding school, my lecturer in mathematics attempted to build the entire calculus in terms of finite elements. It sounded like a good idea at the time: physicists formulate their equations in terms of finite differences—working with finite elements of volume, mass, etc.—and then they take the limit

$$\Delta V \to 0$$

and replace ΔV by the differential dV, etc., getting a differential equation instead of the original finite difference equation. After that, numerical analysts solve this equation by replacing it with an equation in finite differences. The question: "Why bother with the differential equations?" is quite natural. Hence my lecturer bravely started to rebuild calculus, from scratch, in terms of finite differences. Even more brave was his decision to test it on schoolchildren. Have you ever tried to prove, within the ϵ-δ language for limits, the continuity of the function $y = x^{m/n}$ at an arbitrary point x_0 by a *direct explicit computation of δ in terms of ϵ*? The scale of disaster became apparent only when my friends and I started, in reviewing for exams, to actually read the mimeographed lecture notes. We realized very soon that we had stronger feelings about mathematical rigor than our lecturer possibly had (or was prepared to admit, being a very good and practically minded numerical analyst); perhaps my teacher could be excused because it was not possible to squeeze the material into a short lecture course without sacrificing rigor. So we started to recover missing links and to research through books for proofs, etc. The ambitious project deflated, like a pricked balloon, and started to converge to a good traditional calculus course. The sheer excitement of the hunt for another error in the lecture notes still stays with me.

And I learned to love actual infinity—it makes life so much easier.

My story, however, has a deeper methodological aspect. Vladimir Arnold [3] forcefully stated that it is wrong to think about finite difference equations as approximations of differential equations. It is the differential equation which approximates the finite difference laws of physics; it is the result of taking an asymptotic limit at zero. Being an approximation, it is easier to solve and study.

In support to his thesis, Arnold refers to a scene almost everyone has seen: old tires hanging on sea piers to protect boats from bumps. If you control a boat by measuring its speed and distance from the pier and select the acceleration of the boat as a continuous function of the speed and distance, you can come to a complete stop precisely at the wall of the pier, but only after an infinite amount of time: this is an immediate consequence of the uniqueness theorem for solutions of differential equations. To complete the task in a sensible amount of time, you have to allow your boat to gently bump into the pier. The asymptotic at zero is not always an ideal solution in the real world. But it is easier to analyze!

6.3 Proofs by handwaving

> Дано мне тело, что мне делать с ним,
> Таким единым и таким моим?
>
> Osip Mandelstam

After our discussion of the encapsulation of actual infinity, we now return back to lower level cognitive activities.

Mathematical folklore contains a well-known ironic classification of "proofs": proof by blatant assertion, proof by intimidation, etc. Coxeter's proof of Euler's Theorem (Section 2.3) is probably one of the best examples of a *proof by handwaving*. It is fun to tell it in a pub, using fingers as props.

Notice that Euler's Theorem describes the actual movements of a solid body in space, and therefore there are good reasons for handwaving to be relevant for its proof. But mathematics is full of motion and action metaphors applied even to situations where there is no motion or action: we say

- that an asymptote of the hyperbola $xy = 1$ *approaches* the x-axis;
- that a variable t *runs* through the set of real numbers;
- that the sequence $s_n = 1/n$ *converges* to zero as n *tends* to infinity,

and the list of examples can be easily expanded.

Rafael Núñez and George Lakoff ([213], [200]) put the motion metaphor at the center of the cognitive understanding of mathematics. Núñez's paper [213] roots mathematical concepts into unconscious mechanisms linking speech and gesture: the internal mental image of the concept of "convergence" is the shared component of the meaning of the spoken phrase which describes convergence, and a semi-conscious or unconscious gesture which accompanies the phrase.

At first glance, the idea appears to be far-fetched, until one checks the list of experimental evidence showing the intimate relation between speech and gesture. I quote Núñez [213, p. 174], where the list is supported by detailed bibliographic references.

- Speech accompanying gesture is universal. This phenomenon is manifested in all cultures around the world.
- Gestures are less monitored than speech, and they are, to a great extent, unconscious. Speakers are often unaware that they are gesturing at all.
- Gestures show an astonishing synchronicity with speech. They are manifested in a millisecond-precise synchronicity, in patterns which are specific to a given language.
- Gestures can be produced without the presence of interlocutors. Studies of people gesturing while talking on the telephone, or in monologues, and studies of conversations among congenitally blind subjects have shown that there is no need of visible interlocutors for people to gesture.
- Gestures are co-processed with speech. Studies show that stutterers stutter in gesture too and that impending hand gestures interrupt speech production. I have once witnessed how an (admittedly, highly skilled) speech therapist made a severely stuttering child speak flawlessly from the first seconds of the very first speech therapy session: she held the boy's hand to feel his spasms and spoke to him, leaving prompts and pauses for his responses at precision timed moments when the hand was relaxed. For observers—including the boy's mother—this looked like a miracle.
- Hand signs are affected by the same neurological damage as speech [. . .].
- Gesture and speech develop closely linked. Studies in language acquisition and child development show that speech and gesture develop in parallel.
- Gesture provides complimentary content to speech content. Studies show that speakers synthesize and subsequently cannot distinguish information taken from the two channels.

- Gestures are co-produced with abstract metaphorical thinking. Linguistic metaphorical mappings are paralleled systematically in gesture.

At risk of committing the mortal sin of using introspection as a source of empirical evidence, I have to say that I am very sympathetic to Núñez's ideas: they appear to reflect my everyday experiences in mathematics and my observations of other people thinking and talking about mathematics.

I would even suggest the next step. If we think about a mathematical concept as a shared content of a phrase and a gesture, then an *encapsulation* is a mental image of a *completed* potential gesture, the one which we probably have not made yet, but could and would if we were to complete the action inherent in the pre-encapsulation concept.

I will illustrate this principle in a second, using the concepts of convergence and limit. I quote Núñez [213, p. 179]:

> Formal definitions and axioms in mathematics are themselves created by human ideas [...] and they only capture very limited aspects of the richness of mathematical ideas. Moreover, definitions and axioms often neither formalize nor generalize human everyday concepts. A clear example is provided by the modern definitions of limits and continuity, which were coined after the work by Cauchy, Weierstrass, Dedekind, and others in the 19th century. These definitions are at odds with the inferential organization of natural continuity provided by cognitive mechanisms such as fictive and metaphorical motion. Anyone who has taught calculus to new students can tell how counter-intuitive and hard to understand the epsilon-delta definitions of limits and continuity are (and this is an extremely well-documented fact in the mathematics education literature). The reason is (cognitively) simple. Static epsilon-delta formalisms neither formalize nor generalize the rich human dynamic concepts underlying continuity and "approaching" the location.

I disagree with this statement: human dynamic concepts are rich, but not that rich. If one analyzes them, the most intuitive ones happen to be one-dimensional and closely linked to the concept of order (see Section 4.3). The most intuitive special class of continuous functions of real variables is made of piecewise monotone functions which take all intermediate values. Notice that the definition involves only order—and nothing else.[6]

I am not afraid to take the next step and observe that the "dynamics" is frequently present in intuitive mathematical concepts as dynamics of construction: for example, a sequence of numbers "converges" to something or "tends" to something because we *construct* its terms one by one. As the result, I believe that it is more a rule than an exception that a developed mathematical language is static (at least at first glance); an intuition of motion or action (possibly expressed in an unconscious gesture) was useful when we were constructing the new object; it becomes redundant when we are no longer interested, or do not know, or do not care how the object was constructed. The elimination of handwaving is a major paradigm of a development of a mathematical theory.

I have the impression that, in Núñez's understanding, a potentially infinite sequence is dynamic and intuitive, while the actually infinite set of all its elements is static and therefore counterintuitive. Why should this be? Why should an action or a motion be viewed as intuitive while its result should not be?

To illustrate this point, I suggest to have a look at one of the most general forms of the definition of a limit:

Let T be a topological space, x a point in T, and $\mathcal{N}_x$ the filter of neighborhoods of x. A filter $\mathcal{F}$ on T is said to have x as a limit point if $\mathcal{N}_x \subseteq \mathcal{F}$.

I want to apply this definition to an intuitive example of a sequence

$$1, \frac{1}{2}, \frac{1}{3}, \ldots, \frac{1}{n}, \ldots$$

converging to 0. For readers who do not know what filters are, it will suffice to know that we in effect deal with the following sets and functions:

- the set $\mathbb{N}$ of natural numbers;
- the function $\sigma : \mathbb{N} \to \mathbb{R}, \sigma : n \mapsto 1/n$;
- the set $S = \sigma(\mathbb{N})$ of all elements in the sequence;
- the set $\mathcal{A}$ of all subsets in $\mathbb{N}$ which have a finite complement in $\mathbb{N}$;
- the collection $\sigma(\mathcal{A})$ of images of sets from $\mathcal{A}$ under σ;
- the filter $\mathcal{F}$ of subsets in $\mathbb{R}$ generated by $\sigma(\mathcal{A})$ (so that $\sigma(\mathcal{A})$ forms a basis of the filter $\mathcal{F}$);
- the set $\mathcal{B}$ of intervals $(-1/n, 1/n)$ for all $n \in \mathbb{N}$ (they form a basis of the filter $\mathcal{N}_0$ of neighborhoods of 0 in the topological space $T = \mathbb{R}$).

Admittedly, this is quite a formal and messy set-up; does anything new and useful come out of it?

Let us take a closer look.

To say that our sequence converges to 0 is the same as saying that $\mathcal{N}_0 \subseteq \mathcal{F}$, or, equivalently, that for every set $B \in \mathcal{B}$ there is some set $A \in \mathcal{A}$ such that $\sigma(A) \subseteq B$ (one can notice here the same logical form of expression as in Cauchy's classical ϵ-δ definition of limit).

Please notice an important feature of this definition: we deal with
the set $\mathbb{N}$ of natural numbers without specifying a particular order
on it. In effect, the way how elements of S are listed in our sequence
does not matter any longer! For example, we could list elements of
S as

$$1, \frac{1}{10}, \frac{1}{2}, \frac{1}{100}, \frac{1}{3}, \frac{1}{1000}, \dots,$$

at every step jumping arbitrary far ahead and returning to the beginning of the original sequence; this will not affect the definition
of limit. Or you may wish to think about the sequence as a process
of calculating its terms, but in random order, say,

$$\sigma(1) = 1, \ \ \sigma(3) = \frac{1}{3}, \ \ \sigma(7) = \frac{1}{7}, \ \ \sigma(2) = \frac{1}{2}, \dots.$$

This is why the general definition of limit appears to be static:
we abstracted away any concrete method of constructing our sequence. And this is why the general definition of limit works in the
situations where we have no sensible concept of order or direction.

The discussion of differences between the "dynamic" intuition
of limit and its "static" definition that we had in this section can
be boiled down to a difference in the ways we feel and think about
two basic mathematical concepts. One of them is the most intuitive
concept of mathematics, the ordered set of natural numbers,

$$1 < 2 < 3 < 4 < \dots,$$

and the other one is the so-called *Fréchet filter* on the set $\mathbb{N}$ of natural numbers, the set of all subsets with finite complements:

$$\mathcal{F} = \{X \subseteq \mathbb{N} \mid \mathbb{N} \setminus X \text{ is finite }\}.$$

*Humans have an ability to imagine the
result of a completed action which has
not been started yet—and these jumps
of imagination form the cognitive basis of
encapsulation.*

The ordered set of natural numbers
can be understood and dealt with
within the potential infinity framework; the Fréchet filter really requires thinking about the natural
numbers as actual infinity. It is not
immediately intuitive until we apply
to it the great principle of *abstraction
by irrelevance* and bravely ignore the
fact that the natural numbers have
to be somehow built one by one. After we have made *all* the natural numbers, it does not matter any more where we started and in
what particular order we produced them.

I wholeheartedly agree that our intuition of mathematics very
much relies on motion and action metaphors. However, what is
even more important is that humans have an ability to imagine
the result of a completed action which has not been started yet—a
simple observation expressed in famous words:

But I say unto you, That whosoever looketh on a woman to lust after her hath committed adultery with her already in his heart. (Matthew 5:28)

It is likely that similar jumps of imagination form the cognitive basis of encapsulation of mathematical objects and concepts.

Notes

[1]GENERAL CULTURAL INFLUENCES can make certain mathematical concepts appear self-evident—we shall soon see this again using the example of "optimal strategy", Section 7.9.

[2]Weller et al. [142] is a very interesting paper, an example of the application of Ed Dubinsky's APOS theory of learning mathematics. APOS theory describes how **A**ctions become interiorized into **P**rocesses and then encapsulated as mental **O**bjects, which take their place in more sophisticated cognitive **S**chemas. See [111, 112] for a detailed discussion of APOS as applied to the concept of infinity and David Tall [135] for a review of APOS.

[3]It is worth remembering that "sums" of infinite series and limits of sequences can be meaningfully defined in many ways, depending on the context. For example, I recently wrote a paper on group theory where the sequence

$$1, 0, 1, 0, 1, 0, \ldots$$

had limit $\frac{1}{2}$ (the so-called Cesaro limit) and that made perfect sense in the context of the paper. Also, see Section 5.7, page 110, for a discussion of Euler's summation

$$1 + 2 + 4 + \cdots = -1.$$

[4]MULTISETS. Gian-Carlo Rota initiated the study of *multisets* or *ensembles*, which allow *repeated* elements. The concept appears naturally in computer science and in mathematics. For example, the multiset of multiple roots of the equation $x^2 - 2x + 1 = 0$ is $\{1, 1\}$ and is different from the multiset of roots of the equation $(x - 1)^3 = 0$, which is, of course, $\{1, 1, 1\}$. But every element of the first multiset is an element of the second one, and vice versa; therefore the Volume Principle is no longer applicable. We should always remember that mathematical concepts can be developed in many wildly different ways. If we choose a particular way, we have to have a clear idea why we are doing so.

[5]ORDINAL ω^2. In more rigorous terms, you may think of the ordinal number ω^2 as the set $\mathbb{N}^2$ of pairs of natural numbers (m, n) ordered lexicographically:

$$(k, l) \leqslant (m, n) \Leftrightarrow k < m \text{ or } k = m \text{ and } l \leqslant n.$$

[6]There is a well-established area of research on the boundary of real analysis and mathematical logic—the theory of o-minimal structures—where all (definable) functions of a single variable are exactly piecewise monotone functions which take all intermediate values. In naive terms, these are functions whose graphs can be drawn with a pencil.

Part II

Mathematical Reasoning

7

What Is It That Makes a Mathematician?

I have already spent considerable time discussing the workings of the mathematical brain at the subconscious and semiconscious levels. However, mathematics is done by quite conscious reasoning. It would be useful to look, maybe just briefly, at the interaction between the two levels of the mind. This is a classical topic, as may be seen from the famous book *The Psychology of Invention in the Mathematical Field* by Jacques Hadamard [43].

7.1 Flies and elephants

> *A tacit rite of passage for the mathematician*
> *is the first sleepless night caused by an unsolved problem.*
>
> B. Reznick [129]; quoted from Tony Gardiner [114]

> *Detest it* [a certain difficult mathematics problem]
> *just as much as lewd intercourse;*
> *it can deprive you of all your leisure, your health,*
> *your rest, and the whole happiness of your life.*
> Wolfgang Bolyai (in a letter to his son Janos)

To introduce our discussion of the difficulties involved in a mathematician's work, I start with a parable which might look excessively clinical.

During World War II, Sub-lieutenant Zasetsky received a severe head wound which resulted in persistent brain damage. He was observed over 23 years by Professor Luria, who wrote a famous book [204] based on Zasetsky's diaries (the latter comprise more than 3,000 pages). Yuri Manin, when discussing the nature of proofs in his book *Provable and Unprovable* [383], quotes some really astonishing fragments of Zasetsky's diaries:

And more: "Is the elephant larger than the fly" or "Is the fly larger than the elephant". I understood only that "the fly" is small and "the elephant" is big, but, for some reason, could not find my way through the words and answer the question, is the fly smaller than the elephant, or is it larger. The main trouble was that I could not understand what the words "is larger" refer to—the fly or the elephant.

Discussing this fragment, Manin stresses the complexity of the metalanguage text which describes the faults in the understanding of the primary language. In that particular instance, it could be possibly explained by the fact that Zasetsky is talking about the past. But here is an excerpt written in the present tense:

> ... I again try to recall the meaning of the expressions "the fly is smaller than the elephant" and "the fly is larger than the elephant". I try to think about them, what is the correct way to understand them and what is incorrect. If we permute the words in these expressions, they change their meaning. But they look the same to me, as if nothing changed after the words were swapped. But if you think a bit longer, you notice that permutation changes the meaning of these four words (elephant, fly, smaller, larger). But my brain, my memory after I got my wound, and even now, cannot immediately grasp, what the word "smaller" (or "larger") refers to—to the elephant or to the fly. Even in these four words, there are too many permutations.

Manin uses Zasetsky's tortured account to refute Russell's thesis that even a moron should be able to check the validity of a formal proof presented as a sequence of mechanical inferences. Manin comments that, on the contrary, humans are useless at checking formal proofs.

There is a clear difference between higher-level reasoning and lower-level verification and acceptance of elementary facts ("the fly is smaller than the elephant").

Our capacity for higher-level reasoning is so precious a resource because it is so scarce.

We are useless at checking formal proofs because we actively dislike using our higher-level reasoning facilities for routine actions which should normally be done subconsciously. Our capacity for higher-level reasoning is so precious a resource because it is so scarce: Zasetsky was trying to resolve by conscious and controlled reasoning (information processing rate: about 16 bits per second) a problem which is normally handled by the visual processing modules of our brain (information processing rate: 10,000,000 bits per second). (See a discussion of "bandwidth of consciousness", with references to the original psychological research, in Tor Nørretranders's book [211]. The bit rate tables are on pp. 138 and 143.)

I draw two lessons from Zasetsky's account.

First, when teaching mathematics, we have to remember this miserable number: 16 bits per second for conscious information processing (which is further reduced to 12 bits per second for multiplication of numbers or 3 bits per second for counting objects). Our students will not master a mathematical technique or concept unless much more powerful mechanisms of subconsciousness are engaged. Just compare these two numbers: 16 and 10,000,000!

My fellow mathematician, do you recognize yourself in Zasetsky's self-portrait?

The second lesson is about the emotional side of mathematics. My fellow mathematician, do you recognize yourself in Zasetsky's self-portrait?

I do.

It so happened that, half an hour before I read Manin's book, I spent some time in a conversation with a colleague of mine, Maria do Rosário Pinto, trying to figure out whether a certain matrix corresponded to a linear map $U \to V$ or to the map of dual spaces $U^* \to V^*$ (in a context where we had already switched several times, back and forth, between spaces and their duals the issue was not difficult—it definitely was not—but it was highly confusing).

Many—and some of the brightest— mathematicians are "problem-solving" analogues of gambling addicts and adrenalin junkies.

We were in a typical "fly and elephant" situation; this is why reading Zasetsky's confessions minutes later was like a shock to me. Only after my colleague and I used, unsuccessfully, every trick to resolve the issue at the conceptual or intuitive level did we resort to a formal calculation on paper, which, of course, gave us the answer. But what was remarkable was our very reluctance to do the formal calculation; instead, we were seeking ways of making the choice self-evident, because we felt this would be more valuable to us. A calculation establishes the fact, and its result can be formally reused. On the other hand, making a fact self-evident does not establish its formal validity; it still requires a proof. However, self-evident things can be reused, at the intuitive level, in further mental work (I avoid the term "reasoning" here): they will jump up, at the right times, from the subconscious levels of our minds into the areas controlled by conscious reasoning.

I trust that my fellow mathematicians will also agree that Zasetsky's accounts of his mental torture can be used as an explanation, to a non-mathematician, of why mathematics, as a professional occupation, is so uncomfortable. Mathematicians are sometimes described as living in an ideal world of beauty and harmony. Instead, our world is torn apart by inconsistencies, plagued by *non sequiturs*, and, worst of all, made desolate and empty by missing links between words and between symbols and their referents; we spend our lives patching the cracks in the world. Only when the last crack disappears are we rewarded by brief moments of harmony and joy.

And what do we do then? We start to work on a new problem, descending again into chaos and mental pain.

Maybe this truth is not for public consumption, but many (and some of the brightest) mathematicians are "problem-solving" analogues of gambling addicts and adrenalin junkies. My PhD student once complained to me that she was exhausted because for two weeks she awoke every morning with a clear realization that she had continued to think about a problem in her sleep. She was a *real* mathematician. Where can we find more students like her?

Owl (*Otus Persapiens*) as a fledging chick. Photographer: Tom Maack

7.2 The inner dog

I turn to the most controversial, perhaps, thesis of my book—relations between a mathematician and his subconsciousness.

I start by reiterating that mathematics is a language—in this book, we have already discussed this idea a lot.

I dare to add that the language of mathematics contains a dialect or sub-language for communicating directly with sub- and unconscious modules of our mind.

For example, suppose I say to you: "Imagine a triangle and rotate it around the longest side." It is very likely that you will be able to report back to me: "Yes, the resulting volume of revolution is convex and consists of two circular cones with a common basis", or something to that effect. That means that you were able to pass the command to the visual processing centers of your brain, which then managed to unambiguously interpret it and return to you the result in the form ready for verbalization and communicating back to me.

For your subconscious, the language of mathematics is not a natural language; the subconscious has to be trained to understand you.

It is like training a dog.

Dogs have many faculties which we, humans, are lacking—for example, a fantastic sense of smell. To exploit these faculties, we have to send our commands to the dog and interpret its reactions.

A mathematician is a dog trainer; his subconscious is his "inner dog", a wordless creature with fantastic abilities, for example, for image processing or for parsing of symbolic input. A mathematician has to train his "inner dog".

Dennis Lomas,
aged 14

But the art of dog training has to be learned. Learning and teaching dog training are social activities; dog training and handling are also social activities that are rooted in social and economic practice (shepherd dogs, guard dogs, sniffer dogs,...). People have passed to each other optimal recipes for dog handling; the language of dog commands was refined and optimized over the centuries and is now very much standardized and canonized. Interestingly, the Russian for "Bite!" is "Фас!", which is in effect the German "Fass!". A German word was adopted because Russian equivalents were too long—but dogs needed short, clear, distinctive commands.

Perhaps we have to talk about not just one inner dog, but a whole pack—a visual processing dog, say, or a parser dog.

Writing specifically about visualization, Dennis Lomas left the following comment in my blog:

> Your stance seems to be systematically at odds with the traditional picture of mathematics which grants visualization no real significance.
>
> Once visualization is granted such importance, a range of philosophical issues seems to be posed. A philosopher might consider these questions (among others): Are concepts involved in visualization or is visualization non-conceptual? If it is the latter, how can it participate in any justification of a proposition?

He then challenged me with the following quote from Paul Bernays:

> What is special about geometry is the phenomenological character of its laws, and hence the significant rôle played by intuition. Wittgenstein points to this aspect only in passing: Imagination tells us. And this is where the truth lies; one has only to understand it aright (p. 8). The term "imagination"is very general, and what is said at the end of the second sentence is a qualification which shows that the author feels the theme of intuition to be a very ticklish one. In fact it is very difficult to characterize the epistemological rôle of intuition. The sharp separation of intuition and concept, as it occurs in Kantian philosophy, does not appear on

closer examination to be justified. In considering geometrical thinking in particular it is difficult to distinguish clearly the share of intuition from that of conceptuality, since we find here a formation of concepts guided so to speak by intuition, which in the sharpness of its intentions goes beyond what is in the proper sense intuitively evident, but which separated from intuition has not its proper content. It is strange that Wittgenstein assigns no specific epistemological rôle to intuition although his thinking is dominated by the visual. A proof is for him always a picture. At one time he gives a mere figure as an example of a geometrical proof. It is also striking that he never talks about the intuitive evidence of topological facts, such as for instance the fact that the surface of a sphere divides the (remaining) space into an inner and outer part in such a way that the curve joining up an inside point with an outside point always passes over a point on the surface of the sphere. [10, p. 518]

It is clear that Bernays does not separate intuition and concept, and I am in total agreement with him—at least in the visual mode of thinking about mathematics, intuition and concept are almost the same thing. Indeed, a concept (like a "triangle" in my example) is just a command for our "visual dog", the true possessor of our visual intuition.

It is obvious to every working mathematician that, in the professional research community, mathematicians are ranked by the size and strength of their inner dog. Some of them, actually, have inner wolves rather than dogs.[1]

In this book, I am at least trying to write about the "inner dog" of a mathematician, about actual dog commands and my experience of communicating with my own inner dog, about differences in dog command languages for sniffer dogs and guard dogs, etc. There is not much that I can say about the inner workings of a dog—I am not a neuroscientist. But I love dogs and I believe I know how to communicate with them.

Maria do Rosário
Pinto,
aged 7

7.3 Reification on purpose

I wish to make a brief (and very incomplete) list of mental traits necessary for a working mathematician. Of course, obsessive persistence, Zasetsky-style, to retie the torn bonds between concepts should feature prominently on any such list. Friendly relations with the "inner dogs" of subconsciousness are also crucially important, but, at the present stage of development of neuroscience, we cannot describe interactions between mathematicians and their "dogs" in any sensible detail.

Here, I want to return to the discussion of the key element of mathematical practice, reification, and all its ensuing difficulties (Section 6.1), and to add another item to my roster:

> A mathematician is someone who reifies abstract concepts intentionally and purposely and who can reuse, in compressed form, the psychological experience of previous reifications.

It is my conjecture that potential future mathematicians are boys and girls who, at the age when their classmates struggle to reify the concept of a linear equation, can already reify at will (of course, within the limits of the mathematics they know).

A brief case study will possibly be useful. Here is a problem I liked to give at the selection interviews for the Novosibirsk Summer School (the penultimate step of the selection procedure of *Fizmatshkola*, the preparatory Boarding School of the Novosibirsk State University, of which I am a proud alumnus):

Given 2010 distinct points on the plane, prove that there exists a straight line which divides the points into two groups of 1005 points each. (Well, it was in 1978 that I first used this problem, and the problem, correspondingly, was about 1978 points ... .) [?]

Solve the problem without reading further!

I encouraged my interviewees (14–15-year-old boys and girls) to talk about any ideas they could propose towards the solution; I watched attentively for any signs of understanding, on their part, that

(a) the fact deserves a rigorous proof and can be proven; and
(b) the words "there exists" in the problem are likely to mean an invitation to produce an explicit procedure for constructing such a line.

There is at least one simple solution: draw lines through each pair of points; take a line far away from the points and such that it is not parallel to any of the lines through pairs of points; move this new line towards the points, keeping it parallel to its original position. In that way, the moving line will meet the points one by one, thus allowing for counting. All solutions actually produced by children involved similar counting procedures, with lines rotating, circles expanding, etc.

The reader would probably agree that what is required here is the ability to think about the procedure as a single entity, as an object, specifying first the list of requirements for the procedure, and to do that in a one-off problem (most likely, my interviewees never before in their lives encountered problems in any way similar to that one), without the guiding hand of the teacher, without a long series of preparatory exercises.

Reification is difficult; as Anna Sfard describes it [130],

Anna Sfard,
aged 7

> The main source of this inherent difficulty is what I once called the (vicious) circle of reification—an apparent discrepancy between two conditions which seem necessary for a new mathematical object to be born. On one hand, reification should precede any mention of higher-level manipulations on the concept in question. Indeed, as long as a lower-level object (e.g. a function) is not available, the higher-level process (e.g. combining functions) cannot be performed for the lack of an input. On the other hand, before a real need arises for regarding the lower-level process (here: the computational procedure underlying the function) as legitimate objects, the student may lack the motivation for constructing the new intangible "thing." Thus, higher-level processes are a precondition for a lower-level reification—and vice versa! It is definitely not easy to get out of this tangle.

Therefore I add to my list of a mathematician's traits:

> A mathematician recognizes the vicious circle described by Sfard and actively seeks ways to break it.

Surprisingly, the general population contains a number of children who have somehow developed this ability. It is worth mentioning that, in the selection to *Fizmatshkola*, the interviewers were instructed never to ask questions about the children's academic performance at school; the standard interview form filled at each interview contained no fields for the interviewee's school grades. However, we dutifully collected the names of mathematics teachers. As you might expect, a small number of teachers produced a disproportionate number of able students. What always interested me was how these teachers taught; what made their students so special? How can the skill of *reification on demand* be taught?

7.4 Plato vs. Sfard

I wish to briefly visit the realm of philosophy (see more on that in Chapter 12). The concept of reification is (or would be) quite alien to certain philosophical schools. Here is a quotation from Plato:

> [The] science [of geometry] is in direct contradiction with the language employed by its adepts.... . Their language is most ludicrous... for they speak as if they were doing something and as if all their words were directed toward action.... [They talk] of squaring and applying and adding and the like... whereas in fact the real object of the entire subject is ... knowledge ... of what eternally exists, not of anything that comes to be this or that at some time and ceases to be.
>
> [Quoted from Shapiro [88, p. 7]; he refers to [76, 572a]]

The solutions to the 2010 points problem produced by children during my interviews would be unacceptable to Plato, but perhaps quite acceptable to many classical Greek geometers (for example, Hippocrates c. 470 – 410 BC) who used the method of "verging"—sliding of a marked ruler—in geometric constructions and successfully solved the problem of angle trisection. Indeed, the solutions were entirely based on descriptions of *action*, manipulation of geometric entities as if they were real-life objects. Their proofs would not be viewed as rigorous by Hilbert, who, instead of using the word "drawing", famously said in his axioms for geometry (*Grundlangen der Geometrie*, 1899)

> for any two points, there exists a line ... [88, p. 8].

On the other hand, Euclid, and geometers for 2000 years after him, did not mention explicitly the "betweenness" properties of points and lines on the plane, relying instead on the intrinsic mathematical algorithms of their brains which handled "betweenness" with remarkable efficiency. (See the discussion of "betweenness", and its history, with extensive bibliographic references, in Coxeter [296, Section 12.2].) What would Euclid do if challenged with the 2010 points problem? Would he and his followers repair the gap in which they lost the concept of "betweenness"?

7.5 Multiple representation and de-encapsulation

The notion of cryptomorphism (see Section 4.1) and related concepts are rarely discussed in the literature on mathematical education; in conversation with teachers, one can feel that they find the possibility of expressing the same mathematical concept or fact in many different ways more of an obstacle than an advantage. This point of view is reflected in a dialogue of Thompson and Sfard on the nature of reification [130].

Thompson quotes his earlier paper [138, p. 39]:

It is significant that the inventor of algebraic symbolism, Viète, was a professional cryptographer.

I believe that the idea of multiple representations, as currently construed, has not been carefully thought out, and the primary construct needing explication is the very idea of representation. Tables, graphs, and expressions might be multiple representations of functions for us, but I have seen no evidence that they are multiple representations of anything to students. In fact, I am now unconvinced that they are multiple representations even to us...

Pat Thompson later adds that background motivation for this statement "was largely pedagogical". On the contrary, Sfard sees the whole point in that

being able to make smooth transitions between different representations [...] means there is something that unifies these representations.

What she calls a "mathematical object" is such a unifying entity. However, for Anna Sfard mathematical objects are reified mathematical processes, the unifying entities of something the learner of mathematics has already done.

Similarly to what I said in Section 6.1, the crucial difference between a mathematician and a learner of mathematics is that:

> The mathematician actively seeks *new* or known, but previously ignored, representations and interpretations of his or her objects.

The "multiple representation" of Sfard and Thompson is a primitive and passive pre-reification form of cryptomorphism. It is important to stress that the concept of cryptomorphism assumes the *proactive* position of the researcher, a preparedness to recognize the old object appearing, in disguise, in a completely new setting; it includes an element of challenge. At the psychological level, the "crypto" part of "cryptomorphism" emphasizes active problem solving, breaking the code.

It is highly significant that the inventor of algebraic symbolism, Francois Viète, was a professional cryptographer—he served King Henry IV of France.

And here is another problem which I used in my mathematical interviews.

Some anglers caught some fish. It is known that no one caught more than 20 fish and that a_1 anglers caught at least 1 fish, a_2 anglers caught at least 2 fish, and so on, with a_{20} anglers catching 20 fish. How many fish did the anglers catch between them? (Of course, in more concrete versions of the problem $\{a_i\}$ can be replaced by any non-increasing sequence of non-negative integers.) [?]

Solve it!

I remember this problem circulating among my friends in the Summer School. I also remember two principal types of solutions.

Solution 1

Notice that exactly $a_{19} - a_{20}$ anglers caught 19 fish, $a_{18} - a_{19}$ anglers caught 18 fish, and so on. Therefore the total number of fish is

$$20a_{20} + 19(a_{19} - a_{20}) + 18(a_{18} - a_{19}) + \cdots + 2(a_2 - a_3) + (a_1 - a_2),$$

which simplifies to

$$a_1 + a_2 + \cdots + a_{19} + a_{20}.$$

As you can see, some skill in the formal manipulation of sequences would be quite handy.

Solution 2

This solution is more interesting in the context of "multiple representation", since it involves a clear understanding on the part of the solver of the concept of functional dependence and the preparedness to look at one of the most primitive forms of the representation of functional dependence: charts.

Visualization is too intimate a component of mathematical thinking to be entrusted to a computer.

The chart in Figure 7.1 is the best I could squeeze from MICROSOFT EXCEL and is a good example of why software-based learning of mathematics is intrinsically flawed: the software forces on you a particular mode of visualization. However, visualization

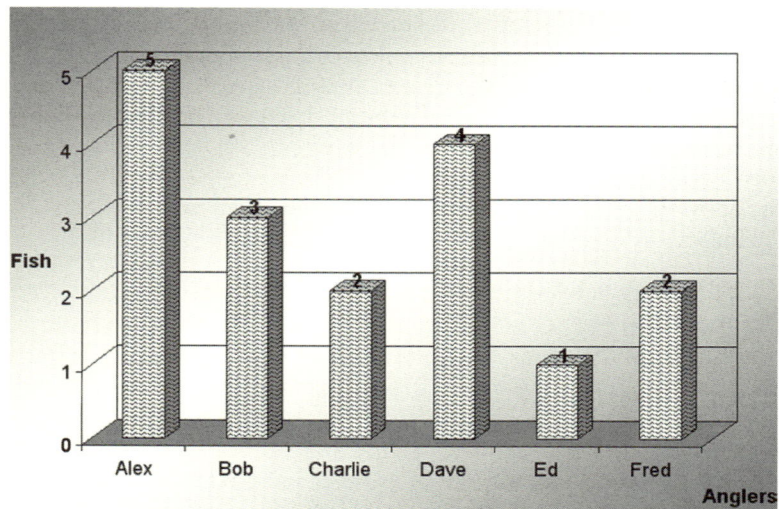

Fig. 7.1. Some anglers caught some fish, a chart made using MICROSOFT EXCEL.

is too intimate a component of mathematical thinking to be entrusted to a computer. Instead, I drew a simple diagram represent-

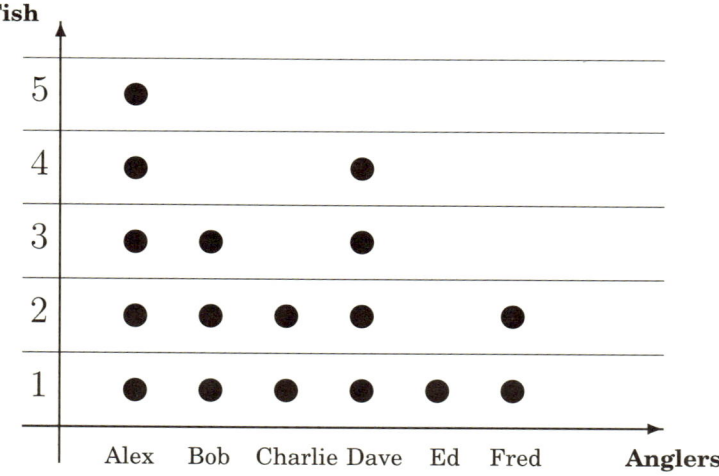

Fig. 7.2. Some anglers caught some fish, represented by cherry stones on graph paper.

ing my own visualization of the problem, the same way I visualized it when I was a schoolboy; see Figure 7.2. I do not know why, but it so happened that I thought about the fish as cherry stones placed on graph paper (at that time, I did not draw an actual picture, but I definitely remember thinking about cherry stones). In this much more primitive chart, a_i is the number of stones in row i, which immediately gives the total number of stones as $a_1 + a_2 + \cdots$.

Here are some lessons to be learned from this small case study:

- The second solution requires a reasonable level of handling of the general concept of a function (of nominal, not numeric, variables: the arguments of the function are names, not numbers!).
- However, the solver should be prepared to choose a very low-level concrete representation of the general concept, or de-encapsulate it down to a rather primitive level. From the teacher's perspective, this means that earlier, lower-level material should be not only well understood by a student, it should be absorbed, interiorized to the point of totally automatic, subconscious use. De-encapsulation is no less important than encapsulation; the student has mastered the encapsulated concept only if she can de-encapsulate it at will and freely choose the most appropriate of many possible modes of de-encapsulation.
- The solver has to actively probe her mind for various representations of the problem (or translations to various languages) until the most appropriate one is found.
- Finally, the problem itself is not that naive: its solution with cherry stones is a miniature version of the Fubini Theorem for the Lebesgue integral. I would bet that the mathematician who originally set the problem knew this connection, and in the most explicit terms. At the personal level, I myself, as an undergraduate student, was overwhelmed by emotion when I recognized my cherry stones in theorems from the course in analysis.
- The first solution of the problem requires a higher level of symbolic mathematical technique. I leave it to the reader as an exercise to find and prove the calculus version of the formula

$$\text{Fish total} = 20a_{20} + 19(a_{19} - a_{20}) + \cdots + 2(a_2 - a_3) + (a_1 - a_2).$$

If you wish to avoid the use of Lebesgue measure, assume that anglers are points on $[0, 1]$ and that the number $f(x)$ of fish caught by angler x is a continuous strictly decreasing function.

[?]   *Do it!*

7.5.1 Rearrangement of brackets

The story with anglers and fish came to my mind recently, when I overheard the following exchange at a meeting of a study group on Kontsevich's motivic integration (in which we attempted to read papers like Hales [361] and Denef–Loeser [343, 344]):

> – I cannot understand why motivic integration has turned out to be so useful. It is nothing more than a very complicated rearrangement of brackets!
> To which my colleague Peter Symonds responded:
> – Well, a lot of mathematics is just a rearrangement of brackets.[2]

The vertical unity of mathematics is not frequently discussed—although it is highly relevant to the very essence of mathematical education.

Many eloquent speeches have been made and many beautiful books have been written in explanation and praise of the incomprehensible unity of mathematics. In most cases, the unity was described as a cross-disciplinary interaction, with the same ideas being fruitful in seemingly different mathematical disciplines and the techniques of one discipline being applied to another. The *vertical* unity of mathematics, with many simple ideas and tricks working both at the most elementary and at rather sophisticated levels, is not so frequently discussed—although it appears to be highly relevant to mathematical education.

7.6 The Economy Principle

The following informal concepts of mathematical practice cry out to be explicated:

beautiful, natural, deep, trivial, "right", difficult, genuinely, explanatory . . .

Timothy Gowers

Quite a number of the phenomena of mathematical practice can be explained in terms of what I call, for lack of a better name, the *Economy Principle*:

> A mathematician has an instinctive tendency to favor objects, processes, and rules with the simplest possible descriptions or formulations.

To some extent the economy is a general tendency of the human mind; it is taken for granted, for example, by composers of IQ tests, where answers to various problems of the type

continue the following sequence: 1, 3, 6, 10, 15

are expected to be based on the assumptions (which incidentally are never stated) that

Peter Symonds,
aged 5

(a) the numbers or objects in the sequence are supposed to be built consecutively one by one, and
(b) the rule for construction has to be as simple as possible.

In that particular case, one can easily observe that consecutive increments in the sequence are 2, 3, 4, 5 and therefore a likely continuation of the sequence is

$$1, 3, 6, 10, 15, 21, 28, \ldots .$$

Number sequences in IQ tests provide some of the best mathematical entertainment on the Internet: use GOOGLE to find an IQ test, copy a sequence and paste it in the search engine of N. J. A. Sloane's *On-Line Encyclopedia of Integer Sequences* [442]. Then look in awe at the astonishing number of mathematically meaningful descriptions and ways to continue the sequence. Notice that each of the sequences has actually appeared in some mathematical problem— the *Encyclopedia* provides comprehensive references! For example, one of the ways to continue our sequence 1, 3, 6, 10, 15 is

$$1, 3, 6, 10, 15, 20, 27, 34, 42, \ldots$$

(sequence A047800). It has the explanation that its n-th term is the number of different values of $i^2 + j^2$ for i and j running through the integers in the interval $[1, n]$.

However, we instinctively know that the first answer is "right" because its description is simpler. Our intuition is reflected in an important concept of modern computer science: *Kolmogorov complexity*. Kolmogorov complexity of a word (or a sequence), as Kolmogorov himself has defined it, is the length of the shortest program producing that word.

We can easily see that the "Economy Principle" is an important part of many informal concepts of mathematical practice. For example, in his talk at the *Mathematical Knowledge* conference in 2004 in Cambridge, Timothy Gowers observed that, in his opinion, a "comprehensible" proof is not necessarily the shortest one, but a proof of small *width*. Here, width measures how much you must hold in your head

François Loeser,
aged 4

at any one time. Alternatively, imagine that you write a detailed proof on a blackboard, carefully referring to all intermediate steps. However, if you know that a certain formula or
lemma will never be used again, you erase it and reuse the space.
A "small width" proof is a proof which never expands beyond one
(small) blackboard.

Igor Pak,
aged 4

In his classical study of the psychology of the mathematical abilities in children [122], Vadim Krutetskii emphasizes
that striving for clarity, simplicity, and economy in a solution
is one of the most important signs of mathematical ability in
children. I quote two examples from his work.

In the first example [122, p. 285], S. G., an eighth grader
(that is, 14 or 15 years old), solves the following problem:

> (*Problem XIX-A-11* of [122]) Find a four-digit num-
> ber with the following conditions: the product of the
> extreme digits is equal to 40; the product of the mid-
> dle digits is 28; the thousands digit is as much less
> than the units digit as the hundreds digit is less than
> the tens digit; and if 3,267 is added to the unknown
> number, the digits of the number are reversed.

It is interesting that S. G. made an explicit choice between
two strategies:

> Initially she composed a complex system of equations in
> four unknowns (the way almost all pupils began). Without
> trying to solve the system she composed, S.G. said: "This
> can be solved but it's very awkward. There ought to be
> a simpler solution here somewhere. But equations aren't
> needed here: 40 can be the product of just two numbers:
> $5 \cdot 8$. But the thousands digit is less than the units digit,
> and then the number is like this: $5 * * 8$. Well, this is clear.
> The number is 5,478."

At this point it also becomes clear why mathematically able children can sometimes be irritating to teachers. When pressed for an
explanation, S. G. clarified:

> "28 is the product of only two digits: $4 \cdot 7$. The hundreds
> digit is less than the tens digit. These digits only have to be
> arranged."

The idea for handling an intermediate step, the uniqueness of
factorization of 40 into digits, $40 = 5 \cdot 8$, was immediately reused by
S. G. as something obvious and not deserving further mentioning.[3]

> A mathematician has an instinctive tendency to
> compress and reuse his/her mental work.

In the second example [122, p. 284], the interviewee is 9-year-old Sonya L.

Problem. A father and his son are workers, and they walk from home to the plant. The father covers the distance in 40 minutes, the son in 30 minutes. In how many minutes will the son overtake the father if the latter leaves home 5 minutes earlier than the son?

Usual method of solution [by 12–13-year-old children]: In 1 minute the father covers 1/40 of the way, the son 1/30. The difference in their speed is 1/120. In 5 minutes the farther covers 1/8 of the distance. The son will overtake him in

$$\frac{1}{8} : \frac{1}{120} = 15 \text{ minutes.}$$

Sonya's solution: "The father left 5 minutes earlier than the son; therefore he will arrive 5 minutes later. Then the son will overtake him at exactly halfway, that is, in 15 minutes."

7.7 Hidden symmetries

Krutetskii's case studies, sharply observed and precisely recorded, frequently contain more than he chooses to highlight and comment on. Actually, in the previous example Sonya L. used a trick which deserves to be specifically mentioned: she noticed a hidden symmetry in the set-up of the problem and immediately exploited it.

> A mathematician seeks and exploits hidden sym-
> metries in a problem.

The word "symmetry" here has to be understood in the widest sense and applied not only to geometric symmetry as we know it, but also to "semantic" or "logical" symmetry.

The famous *Pons Asinorum* theorem[4] of Euclidean geometry provides a very poignant example. The theorem follows:

the base angles of an isosceles triangle are equal.

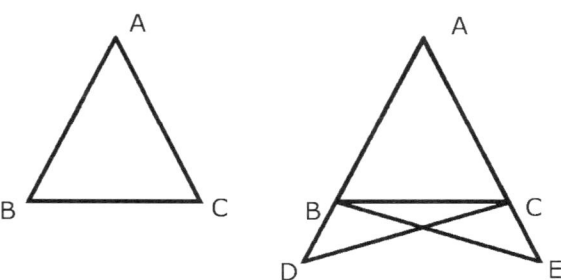

Fig. 7.3. *Pons Asinorum.* Euclid uses an auxiliary construction to avoid application of the Side-Angle-Side Criterion of congruence to the same triangle. For that, he extends the sides and takes points D and E so that $AD = AE$. Then we shows $\triangle DAC = \triangle EAB$ by the Side-Angle-Side Criterion and at the next step that $\triangle CBD = \triangle BCE$. So $\angle CBD = \angle BCE$. After that, $\angle CBA = \angle BCA$ as exterior angles to equal angles.

My own teaching experience (back in Russia in the 1980s) showed that surprisingly many students were able to see that *Pons Asinorum* could be proven by a direct argument based on the *formal* symmetry of the premises:

1. $AB = AC$ and $AC = AB$
2. $\angle BAC = \angle CAB$ (since the angle is equal to itself).
3. $\triangle BAC = \triangle CAB$ (by the Side-Angle-Side Criterion of congruence).
4. Therefore, $\angle B = \angle C$.

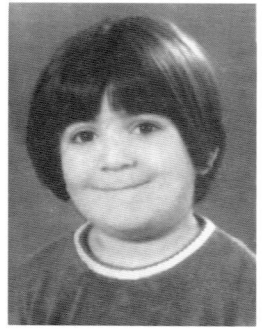

Ayşe Berkman,
aged 6

The proof in the school textbooks was, of course, different because the "formal symmetry" proof was deemed to be too difficult for schoolchildren (and it probably was for many of them). Instead, the *Pons Asinorum* was proven outside the axiomatic system, by a direct application of the bilateral symmetry of the triangle viewed as a cardboard cutout. This approach was promoted by Hadamard in his highly influential *Leçons de géométrie élementaire* [297] and adopted by canonical Russian textbooks.

The dispute about the usability of the "formal symmetry" proof in teaching apparently has a long and honorable history—perhaps starting with Euclid himself, who did not use it in his *Elements*.[5] With the advent of computers and Artificial Intelligence the story found a fascinating turn—proofs based on "semantic symmetry" (the term is from [354]) happened to be natural for automated proof systems. The first breakthrough was made by a famous computer scientist, Marvin Minsky: in 1956, his (hand-simulated) program easily found the "formal symmetry" proof of *Pons Asinorum*. A comment by Minsky is quite revealing:

> What was interesting is that this was found after a very short search—because, after all, there weren't many things to do. You might say the program was too stupid to do what a person might do, that is, think, "Oh, those are both the same triangle. Surely no good could come from giving it two different names." [447]

As a side comment, I wish to express my regret that proofs are increasingly suppressed in mathematics teaching. I can say from my teaching experience at Manchester that students from Greece and Cyprus handle the concept of proof much better than British students. Euclid happened to be Greek, and, as a matter of national pride, Euclidean geometry is still being taught in Greek schools.

7.8 The game without rules

Perhaps one of the least discussed professional traits of a mathematician is the ability to formalize and explicate still vague ideas and constructions.

> A mathematician can work in an incomplete set-up and recover the formal assumptions and context which makes the problem meaningful.

Remarkably, this trait can be found in young children, in the form of an instinctive ability to make up—and then follow—the "rules of the game" in the course of a game.

Mathematical folklore is full of problems where the set-up is intentionally left incomplete, in the expectation that the solver will be able to recover the rest. Here is my favorite one; it belongs to the genre of *interview problems*. It is best told in a face-to-face chat with a child, with the help of some scrawlings on scratch paper.

THE BEDBUG PROBLEM. A student lives in a dormitory room infested with bedbugs (*Cimex lectularius*). He has bought a new bed and wishes to protect it from infestation. Since bedbugs can crawl over any surface, but cannot swim, the usual solution is to put the legs of the bed into tins with water. However, bedbugs also have a devilish ability to crawl, upside down, over the ceiling, position themselves above the desired target and then fall onto it. Our student made a large tin vessel and hung it over the bed (Figure 7.4). Unfortunately, it does not help. Now we can formulate a problem: assume that the student has an unlimited supply of tin and can make vessels of any size and shape. How can he protect himself from the bedbugs?

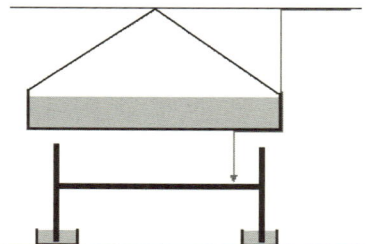

Fig. 7.4. Combining crawling and falling, a bedbug can easily find its way even to a well-protected target.

It is interesting to see how children handle many unstated assumptions of the problem. For example, the student needs to breathe and therefore cannot seal himself hermetically in a tin capsule. Can a bug fall precisely at a very thin edge of the vessel? What happens if the tin is *very* thin? Can a bug crawl over the thin wires from which the vessel is hung? However, the problem has an indisputable solution which immediately removes all these questions (Figure 7.5). Remarkably, I have met many children who, although they have never before seen problems like this, were ready to accept and trust that I, as a problem poser, and they, as problem solvers, follow some undisclosed set of rules which will become immediately clear when the solution is found.

The reader will find more loosely formulated problems in Chapter 10. The cycle of Post Office Problems discussed there presents an even harder challenge: prior to figuring out the rules of the game, the solver has to understand first in what mathematical language he or she has to formulate the rules.

These problems appeal to one of the most natural ways of encapsulating complex processes: treat them as a game, or, more precisely, as *make-believe play*. (For otherwise, why do children, in all cultures, play games?) Crucially, we have to figure out the rules of the game as we play and make the play a consistent and believable whole. When the rules become clear, the psychologically easiest way to accept and follow them is to treat them as the proverbial rules of the game!

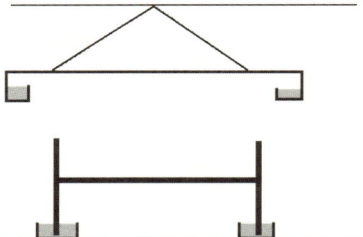

Fig. 7.5. A solution to the Bedbug Problem. As you can see, the thickness of the tin does not really matter.

The game context makes the natural *abstraction by irrelevance* perhaps the simplest and the most powerful form of mathematical abstraction. Lines have no width, not because we want them so, but because we do not care about the width: we are using them in situations where width does not matter—just as we do not care about the thickness of tin in the Bedbug Problem.

7.9 Winning ways

The umbrella term "game" covers a continuous spectrum from "make-believe play," where rules are made on the fly, to established games like chess, where the player has to accept the rules as they are, no matter how strange or bizarre they may be. As mental activities, formally described games come closest to mathematics. It is an observation by G. H. Hardy [45] that chess problems are exercises in combinatorics and that their solutions are instances of mathematical proof in one of its most basic forms: proof by listing all possibilities. In *A Mathematician's Apology*, G. H. Hardy again and again returns to the comparison between mathematics and chess, and, of course, his findings are in favor of mathematics:

> A chess problem also has unexpectedness, and a certain economy; it is essential that the moves should be surprising, and that every piece on the board should play its part. But the aesthetic effect is cumulative. It is essential also (unless the problem is too simple to be really amusing) that the key-move should be followed by a good many variations, each requiring its own individual answer. [...] All this is quite genuine mathematics and has its merits; but it is just that 'proof by enumeration of cases' (and of cases which, at bottom, do not differ at all profoundly) which a real mathematician tends to despise.

And here we come to one of the strangest episodes in the history of mathematics. Unlike games of chance, which attracted the attention of mathematicians from the 17th century onwards and led to the creation of probability theory, deterministic games were ignored by mathematicians for two or three more centuries. The first theorem of game theory—remarkably, about chess—belongs to the famous set theorist Ernst Zermelo and was published in 1913 in his paper *An application of set theory to the theory of chess* [417]. I state the result in its modern formulation; will the reader be surprised by the theorem?

In the game of chess, at least one of the players has a strategy which assures that he either wins or, at least, never loses the game.

Of course, nowadays it is self-evident; but it was not so in the beginning of the 20th century, when, in the mass culture, chess was assumed to be a psychological game, something like poker—although Steinitz had already initiated a scientific approach to the game, building on earlier work by Philidor. Interestingly, Lasker, a prominent mathematician and one of the leading chess players of the beginning of 20th century, was much criticized for his "psychological" style of play.

A clear formulation of results in game theory requires use of sets. A modern commentator [369] writes:

It is a measure of how far set-theoretic thinking has become embedded in mathematics that today we would regard the "application of set theory" in Zermelo's paper as merely set-theoretic formulation in set-theoretic notation. However, it must be remembered that such formulations for the mathematization of problems were quite novel at the time.

Ali Nesin,
aged 15

Actually, Zermelo was interested in more than that. He defined the concept of a "winning" position and asked the question: in a winning position for White, how many moves will it take to reach a checkmate? Using the fact that the game of chess has only finitely many possible positions, he argued that, starting from any winning position, a player can force a win in at most N moves—no matter how the other party plays—for some fixed natural number N.

The concept of "winning strategy" was not explicitly formulated by Zermelo, but, as Dénes König [377] later showed, it was implicitly present in Zermelo's arguments. A *winning strategy* for a player is a function from the set of positions to the set of moves such that if one plays according to this function, he always wins, no matter what the moves of the other player are. Also, embedded in Zermelo's paper was the concept of *determinacy*: the notion that one of the players has a strategy that ensures a win or a draw.[6]

As soon as the concept of winning strategy was explicitly formulated, it became clear that it was remarkably intuitive. In mathematics, its power of encapsulation should not be underestimated.

As an example, let us consider the following problem.

> In the game of "double chess" both players are allowed to make two moves in a row. Prove that White has a strategy which ensures a draw or a win.

A solution is deceptively simple: assume that White has no such strategy. Then Black has a winning strategy, and White, moving a knight forth and back, returns the chessboard into the pre-game state and yields the first moves to Black, in effect, changing his own color to Black. But Black has a winning strategy—a contradiction.

The notion of a "winning strategy" is an example of a mathematical concept which slipped into mass culture without its escape ever being noticed by mathematicians.

This is a pure proof of existence: it says nothing whatsoever about the actual strategy! At this point, it is worth recalling the famous words by G. H. Hardy [45]:

> *Reductio ad absurdum*, which Euclid loved so much, is one of a mathematician's finest weapons. It is a far finer gambit than any chess play: a chess player may offer the sacrifice of a pawn or even a piece, but a mathematician offers the game.

It is interesting to compare the solution of the "double chess" game with other "yield the first move in a symmetric situation" strategies, as in the following game:

> Two players take turns placing equal round coins on a rectangular table. The coins should not touch each other; the player who places the last coin wins (and takes the money). Describe the winning strategy for the first player.

This is a simple game, and the solution is simple: the first player has to place his first coin exactly at the center of the table and then mirror the moves of the second player (under $180°$ rotation with respect to the center of the table). This is a good example of a strategy as a simple rule which prescribes how one has to react to the moves of another player. In the double chess game, the levels of compression and encapsulation are much higher: we have forced White into the ridiculous situation that he must react to the whole optimal strategy of Black—without even knowing whether Black' strategy brings victory or just a draw.

It is not a coincidence that the first-ever theorem of game theory was proven by a set-theorist.

The notion of a "winning strategy" is an example of a mathematical concept which has slipped into mass culture without its escape ever being noticed by mathematicians. I can safely use the "double chess" problem in my class as an example of a *Reductio ad Absurdum* argument without being afraid that the students will notice a subtle flaw in the argument: in our solution, we assume without proof that at least one of the players has a strategy which assures him of at least a draw—that is, Zermelo's Theorem for "double chess". Zermelo's Theorem is self-evident to my students!

However, the determinacy of a game, that is, the existence of a winning strategy for one of the players (without even specifying which one) can be an extremely powerful statement and has to be handled with care. Let us allow ourselves to soar for a second to the rarified altitudes of set theory. The *Axiom of Determinacy* can be added to the Zermelo-Fraenkel axioms of set theory instead of the Axiom of Choice, and it produces a very cosy world where, for example, every set of real numbers is Lebesgue measurable. The Axiom of Determinacy can be formulated in terms of a simple game:

Identify the real segment $[0, 1]$ with the set of all infinite decimal fractions

$$0.a_1 a_2 a_3 \cdots$$

where all the a_i are digits from 0 to 9. Let X be a subset of $[0, 1]$. Two players write, in turn, the digits b_1, b_2, etc.; if the resulting real number $0.b_1 b_2 b_3 \cdots$ belongs to X, the first player wins, and if not, the second player wins.

The Axiom of Determinacy claims that, for every X, one of the players has a winning strategy. ⬚ [?] ⬚

If X is the set of rational numbers in the segment $[0, 1]$, which of the two players has a winning strategy?

Of course it is not a coincidence that the first-ever theorem of game theory was proven by a set-theorist: its rigorous formulation requires the use of set-theoretic concepts. A *strategy* is a function from the set of all possible positions (in the case of chess, a finite, but huge, set) to the set of all possible moves. An optimal strategy is a particular function: it should have the property that, being mechanically applied to every position in the game, it ensures the best outcome of the game, no matter what the moves of the second player are. (If the position is already lost, an optimal strategy should delay the defeat for as long as possible; if it is winning, it should ensure the quickest victory.)

Although the Axiom of Determinacy deals with very sophisticated sets, a strategy is an innocent looking function from the set of finite decimal fractions $\{0.a_1 \cdots a_n\}$ into the set of digits

$\{0, 1, 2, \ldots, 9\}$, or, in simpler terms, a rule for continuing the decimal fraction.

The game of 8×8 checkers has been completely solved: recently, an extensive computer analysis has shown that if two players face off at checkers and neither makes a wrong move, then the game will inevitably end in a draw [399].

Chess is a much more complex game. But even in chess, we already have in our possession a non-trivial chunk of the optimal strategy (of course, solutions to chess problems are also tiny fragments of an optimal strategy). Indeed, chess endgames with a small number of pieces have been exhaustively analyzed with the help of computers. The first breakthrough was made in the early 1980s by Ken Thompson— I remember how astonishing his work was at that time. Thompson generated the list of all legal positions with King and Queen versus King and Rook (quite a long list, about two million positions). Then he programmed a computer to work backwards from mates and compose a table of optimal moves for every position. His table also included the distance to mate—61 moves in the longest game. Before Thompson, this endgame, although believed to be a win for the King and Queen, was known to be extremely difficult if played against a well-trained opponent.[7] Similar "endgame databases" for most six-piece endings are now commercially available.[8] Thompson famously said that his database plays "like God": in every position it knows the outcome of the game. If it is a win, the database knows a move leading to quickest victory, and if a loss, a move leading to the longest defence. When it first appeared, Thompson's table had the most demoralizing impact on human players: many of its moves were completely counterintuitive for a human. [9]

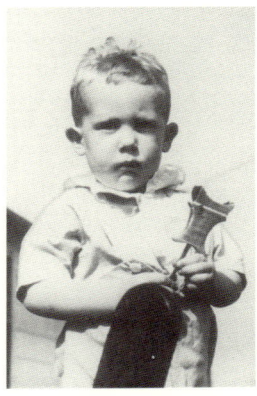

Gregory Cherlin,
after winning
an egg rolling
contest.

Endgame databases provide an example of good computer science, but, in the opinion of many, poor mathematics: just recall the words of G. H. Hardy [45] that non-precisely the *solution* of a chess problem is

> just that 'proof by enumeration of cases' ... which a real mathematician tends to despise.

It would be interesting to trace how the concept of "winning strategy" was absorbed by the mass culture. One would also like to know which had the greater impact on public awareness— computer games or the gradual spread of the ideas of mathematical economics in the form of "portfolio management" and things like that. (Consider the managerial newspeak of "competitiveness" and "optimal solutions".) Game theory poses an awkward question for mathematical education: many of its concepts (say, Nash equilibrium and Pareto efficiency) are simple and can be effectively taught at the secondary school level, with a variety of lively and challenging problems and a host of real-life applications; they may even lead to a discussion of moral issues ("Nash punishment" is a convenient pretext). Whether it provides a fledging mathematical mind with the same level of stimulation as, say, Euclidean geometry is a non-trivial proposition. It would be interesting to explore it—avoiding, of course, non-reversible experiments on children. When I read descriptions of some of the experiments in mathematical education, I wanted to call the hotline of NSPCC (National Society for the Prevention of Cruelty to Children) and report child abuse.

7.10 A dozen problems

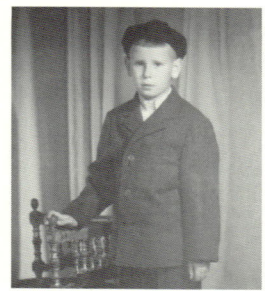

This section contains a sample of archetypal mathematical problems, very elementary, but at the same time reflecting the principal paradigms of mathematical thinking. I wish to give the flavor of what we would expect from high school graduates who might be interested and able to pursue a professional career involving serious mathematics—not necessarily within the narrow domain of mathematics itself.

In that sense, my list is dual to Arnold's famous list of 100 mathematical problems which every mathematics or physics graduate should be able to solve [103].

Of course, the list could be made much longer. A few problems mentioned elsewhere in the text could be added to it, but I prefer to avoid duplication.

Eugene Khukhro,
aged 7

7.10.1 Caveats

However, I feel that I have to make some important caveats.

• These problems are **not** tests of mathematical abilities. If a boy or girl can solve any of them, he or she deserves some attention from the mathematics teacher. Nonetheless a child without previous experience of non-standard problems might still have distinctive mathematical abilities but fail to solve these particular problems.
• If a child is systematically exposed to non-standard problems, he or she quickly expands the range of problems accessible to him/her. Anyone who runs mathematical competitions probably knows children who can solve most of the problems on the list.
• To see how a child copes with a problem, his/her answer or even a detailed solution is not enough. You have to talk to the child

while he/she works on a problem, and, without giving him/her
any hints, trace his/her line of thought.

- The whole spirit of such a list is that it should be as "curriculum independent" as possible. However, in the absence of a well-designed school mathematics curriculum, most potentially able students never develop the skills or attitudes which bring non-standard problems within range. Thus these problems should not be treated as diagnostic.

- The problems in this selection have different levels of "difficulty", and the sample is not as representative as one might wish.

7.10.2 Problems

Problem 1. It takes two hours for Tom and Dick to do a job. Tom and Harry take three hours to do the same job. Dick and Harry take six hours for the job. Prove that Harry is a freeloader.

Problem 2. It takes five days for a steamboat to get from St. Louis to New Orleans, and it takes seven days to return from New Orleans to St. Louis. How long will it take for a raft to drift from St. Louis to New Orleans?

Problem 3. Write all integers from 1 to 60 in a single row:

$$123456789101112131415161718 19 \ldots 5960.$$

From this number, cross out 100 digits so that the remaining number is

(a) the smallest possible;
(b) the largest possible.

Problem 4.
The number

$$1\ 00000\ 00000\ 30000\ 00000\ 00070\ 00000\ 00021$$

is the product of two smaller natural numbers. Find them.

Problem 5.
"Our teacher Mr. Jones has more than 1000 books," said Tom.
"Oh no, he has less than 1000 books," said Gareth.
"Well, Mr. Jones definitely has at least one book," said Helen.

If only one of these statements is true, how many books does Mr. Jones have?

Problem 6. One hundred statements are written in a notebook:

This notebook contains 1 false statement.
This notebook contains 2 false statements.
This notebook contains 3 false statements.
. . .
This notebook contains 100 false statements.

Which of these statements is true?

Problem 7. Here are several dates in Swahili:

tarehe tatu Disemba jumamosi; tarehe pili Aprili jumanne; tarehe nne Aprili jumanne; tarehe tano Octoba jumapili; tarehe tano Octoba jumatatu; tarehe tano Octoba jumatano.

The translations in English are given in random order:

Monday 5 October; Tuesday 2 April; Wednesday 5 October; Sunday 5 October; Saturday 3 December; Tuesday 4 April.

Write in Swahili: Wednesday 3 April; Sunday 2 December; Monday 1 November.

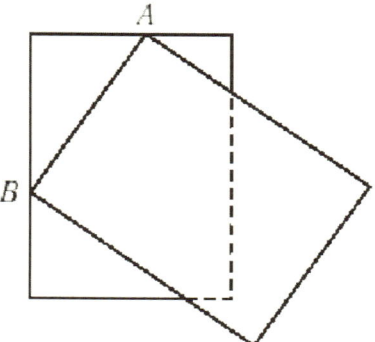

Fig. 7.6. For Problem 8: two sheets of paper.

Problem 8. Two sheets of paper (of the same size) are placed one onto another as shown on Figure 7.6, so that corners A and B of the upper sheet lie on the sides of the bottom sheet, and one corner of the bottom sheet is covered. Which part of the bottom sheet is bigger: that covered by the upper sheet or the part uncovered?

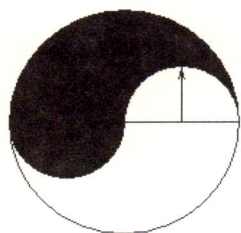

Fig. 7.7. For Problem 9: Yin and Yang.

Problem 9. The shaded region in Figure 7.7 is bounded by three semi-circles. Cut this region into four identical parts, i.e., parts of equal size and shape (but possibly of various orientation) [444] .

Problem 10. A bar of chocolate is subdivided by grooves into 40 segments, arranged in 5 rows and 8 columns. How many times does one have to break the bar to get all 40 segments?

Problem 11. Estimate, approximately, how many solutions the following equation has:

$$x = 100 \sin x.$$

Problem 12. A rectangle of dimensions 19×91 is cut by straight lines parallel to its sides into $19 \times 91 = 1,729$ equal squares of side 1. How many of these squares does a diagonal of the rectangle cross?

7.10.3 Comments

Problem 1.

This is a non-standard modification of problems about "rates", which were a staple of arithmetic textbooks of yesteryear (well, of the early 20th century). Usually they were formulated in terms of pipes filling water tanks. This amusing formulation belongs to Hovik Khudaverdyan.

Problem 2.

This is a non-standard repackaging of the same kind of ideas as in Problem 1. I have met children who can immediately see the similarity: "Ah, yes, it is about one over something and taking the sum, or difference, or whatever... ."

Problem 3.

A successful solver must first explicate the concepts "bigger" and "smaller" as applied to decimal numbers—only after that do logical manipulations become possible.

Problem 4.

The problem engages a logically subtle and elusive connection between "factorization" in algebra and in integer arithmetic. The issue is not so much the algebraic content as the child's ability to seek and see connections between different areas of mathematics.

The problem is taken from a first year university course on number theory and cryptography. When building their own toy implementations of the RSA cryptographic system, students have to produce, with the help of the MATLAB software package, products $n = pq$ of two large prime numbers p and q, sufficiently big so that n cannot be factorized by the standard routines of MATLAB or MATHEMATICA. A surprising number of students end up with numbers like the one in the problem. Even when explicitly prompted, not every student can immediately see the source of potential trouble.

Problem 5.

This is one of the examples which show how difficult it is to draw the line between natural mathematical abilities and the "mathematical culture" absorbed at school. Besides an inclination to do combinatorial logic, the successful solver must have a very clear understanding of the meanings of the expressions "less" and "more". Also, the solution is not unique: it is interesting to see how children react to the problem being just a little bit undetermined.

Problem 6.

This is a trickier case of the logic of self-referential systems of statements; here, they refer to the number of other statements being valid. Notice also the ambiguity of the formulation: the answer depends on whether the statement

> this notebook contains 1 false statement

is understood as

> the notebook contains *exactly* 1 false statement

or

> the notebook contains *at least* 1 false statement.

Problem 7.

Please notice that the number 1 does not appear in the sample English translations. If you think that this makes it impossible to translate the last date, Monday 1 November, try to figure out where the numeral "one" sits in Swahili sentences?

Problem 8.

It is interesting to compare this problem with one of the problems used in Celia Hoyles's and Dietmar Küchemann's study of the development of the concept of proof in schoolchildren [120]; see Figure 7.8.

Celia Hoyles,
aged 7

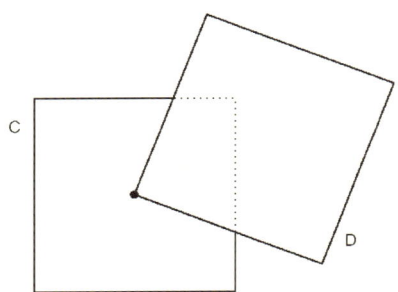

Fig. 7.8. Squares C and D are identical. One corner of D is at the center of C. What fraction of C is overlapped by D? Explain your answer.

As you can see from the comparison, Problem 8 is also about finding hidden symmetries, but in a less symmetrical set-up.

Problem 9.

One's first thought is likely to be that this must be impossible. So the problem tests the flexibility of thinking (since one has been challenged to succeed, so it is clear that one is missing something) and the powers of visualization. Actually, the shaded region can be divided into any number of identical parts.

Problem 10.

The simple counting nature of the problem is intentionally obscured by excessive details.

Problem 11.

This is the only problem on the list which goes beyond the English curriculum up to age 14. Of course, GCSE and A-level mathematics provide much more material for exciting problems of any level of difficulty.

Problem 12.

This problem moves slightly beyond what the English school-children are currently taught. The solver needs to use the fact that 19 and 91 are relatively prime—yet "common factors" and "relative primeness" are treated only in the context of fractions.

The problem is interesting for the dynamics of encapsulation/de-encapsulation: to solve the problem, you have to remember that numbers are the results of counting processes (this also applies to Problem 10) and you have to understand how a dynamic process fits into a static geometric picture.

Notes

[1]I mention in passing that one of my colleagues referenced in this book grew up in Siberia in a kind of home where his folks kept, in place of a guard dog, a wolf on a chain. The little boy and his wolf were best friends—the boy hugged the wolf, fed him from his hands, etc. And, as a mathematician, my friend could be best described as a mature alpha wolf.

[2]EVEN MORE ABOUT REARRANGEMENT OF BRACKETS. I wish to add a few more words about "rearrangement of brackets", this time in the context of social studies. Mathematical economics frequently uses exceptionally simple mathematical ideas. When a study is done with a political effect in mind, the intentional simplicity of the mathematics used is a bonus. Mathematics is still viewed as the embodiment of truth, and its potential moral impact should not be underestimated.

One of my favorite examples is the *Blinder-Oaxaca decomposition* introduced by Blinder and Oaxaca in the 1970s in the context of the Equal Rights Movement [316, 390]. The so-called linear regression analysis of statistical data allows one to derive the wage equation

$$W = CE$$

where W is the wage, E is the vector of parameters describing the person's experience (its components usually include years in education, years in employment, etc.), and C is the vector of coefficients derived from the statistical data. Now look at the wage data for men and women and derive the wage equations for men and women separately:

$$W_m = C_m E_m, \qquad W_f = C_f E_f.$$

It is a well-known property of linear regression that if

$$W_m, \ W_f, \ E_m, \ E_f$$

are sample means, then these equations hold exactly.

Now recombine the two equations expressing the wage gap $W_m - W_f$:

$$\begin{aligned} W_m - W_f &= C_m E_m - C_f E_f \\ &= C_m(E_m - E_F) + (C_m - C_f)E_f. \end{aligned}$$

In this expression, the first term $C_m(E_m - E_F)$ is the part of the wage gap due to differences in average characteristics between men and women (explained or non-discriminatory part). In particular, if men and women had the same levels of education, experience, etc., these terms would be 0. If education, experience, etc., all raise wages (that is, all components of C_m and C_f are positive) and if men are better educated, have more experience, etc., then a positive share of the gap is explained.

Much more interesting is the second term, $(C_m - C_f)E_f$, the discriminatory part of the age gap: it is due to differences in treatment of men and women in the labor market.

In his PhD thesis in the 1970s, Oaxaca used this decomposition in his classical study of labor discrimination of black people in USA, and his findings were striking. The simplicity of the underlying algebra very much helped to bring his results into the mainstream political discourse.

But as frequently happens in statistics, as soon as one starts estimating the statistical errors inherent in the estimated values of coefficients and their subsequent impact on the precision of the gap decompositions, things become much more complicated. A survey of a considerable body of work can be found in [391].

[3]Reuse of mental work is recognized, in the mathematical folklore, as one of the easy-to-ridicule aspects of mathematics, e.g., how to boil a pot of water sitting one foot north of a fire—displace it one foot south and wait—and how to boil a pot of water sitting one foot south of a fire—displace it two feet to the north, arriving at a known case. This joke is contributed by Gregory Cherlin.

[4]PONS ASINORUM. Coxeter writes in [296]: "The name *Pons Asinorum* for this famous theorem probably arose from the bridge-like appearance of Euclid's figure (with the construction lines required in his rather complicated proof) and from the notion that anyone unable to cross this bridge must be an ass. Fortunately, a far simpler proof was supplied by Pappus of Alexandria about 340 A.D." [And that was exactly the "formal symmetry" proof which I discuss in this book.– AB]

[5]MORE ON PONS ASINORUM. Notice that Euclid, in his proof of the theorem (Figure 7.3), accepts that an angle can be congruent (equal) to itself—as a *common* angle of two triangles—but stops

short of extending reflexivity of congruence to triangles. The identity of mathematical objects is a delicate issue.

[6]The reader interested in the history of game theory may wish to consult [86] for a discussion of Zermelo's paper and related early papers on game theory, including an expansion of Zermelo's Theorem by Lázló Kalmár [368].

[7]In tournament play, after reaching Queen vs. Rook, the weaker side normally resigns. But there are cases at the grand master level in which the Rook player has insisted on continuing and has held off his opponent long enough to draw.

[8]See Ken Thompson's paper [411] about the analysis of 5-piece endgames in chess and the book *More Games of No Chance* [389] for the later results in the analysis of endgames in various other games.

[9]CHESS. Gregory Cherlin made an interesting comment on the difference between mathematicians' and chess players' approaches to the analysis of endgames:

> However, the analysis is given only for the 8×8 board and I have never understood whether the general problem is a win or a draw. One tends to use very precisely the narrowness of the board when forcing the King back in the early stages.
>
> Doron Zeilberger has a student who has been investigating optimal strategies for simple endgames on an $n \times n$ board, looking for the minimal length of a forced win as a function of n. I mentioned the Queen-Rook endgame but I don't know if it was pursued.
>
> It seems to me that there is a good deal of concept formation possible in this kind of simple endgame and that it is not understood mainly because the experts, who know a great deal and have some useful concepts, are not interested in formulating concepts for problems with known solutions on the standard board. They speak of Rook pawns, Knight pawns, Bishop pawns, and center pawns, each with different properties. But I think on a larger board one has the same four categories, namely three exceptional wing pawns and then the generic case. On the 8×8 board the general notions occur less often than the exceptions ... but they do occur. Most of endgame theory would be clarified by doing it on an $m \times n$ board. In particular the Russians have a number of diagrams showing the precise location of a key piece needed to force a win (the King, or possibly a Rook). These regions consist mostly of irregularities on a small board but on a larger one presumably have the same number of exceptional points on top of a very simple geometry.
>
> Of course, to make this worth looking at one has to begin by making it useless.
>
> This is an astonishingly undeveloped field, in spite of the many years expended and encyclopedias written, and which does not require a computer for research.

8

"Kolmogorov's Logic" and Heuristic Reasoning

This chapter is dominated by the "vertical integration" thread of my narrative: I look at two examples of the application of some basic heuristic principles of invention in mathematics. They yield remarkable and unexpected results but, at the same time, are based on very simple mathematics.

I start by looking at a parallel example (and a paradigm) from the general area of technical invention and briefly recount one of the most fascinating stories from the history of technology in the 20th century: Hedy Lamarr and her contribution to spread-spectrum communication. Then I look at mathematical structures for sonar signals and continue my discussion of the problem of dividing 10 apples among 5 people (Section 4.7), this time on somewhat different material: turbulence in the motion of a fluid.

8.1 Hedy Lamarr: a legend from the golden era of moving pictures

> *Any girl can be glamorous.*
> *All you have to do is stand still and look stupid.*
>
> Hedy Lamarr
>
> *Films have a certain place in a certain time period.*
> *Technology is forever.*
>
> Hedy Lamarr

Unusual for a book on mathematics, I have to start this chapter by briefly recapping the life story of a Hollywood star of yesteryear, Hedy Lamarr. Everything that I know about her I picked on the Internet, and I have to warn the reader that discerning myth from reality was not my priority. Therefore I feel that I have to be as brief as possible.

Hedy Lamarr, née Hedwig Eva Maria Kiesler, was born in Vienna in 1914 in a cultured bourgeois family. Her father was a bank director, her mother a pianist. Family connections in the artistic world led to Hedy's acting talents being discovered early. She achieved international fame and cult status in 1933 for her role in the Czech film *Ecstasy*. By the standards of her time, the role was *risqué* for an 18-year-old girl from a good family. The film caused a considerable controversy—which, of course, only added to Hedy's fame.

In the same year, Hedy Kiesler married the Austrian industrialist Fritz Mandl, Director of the *Hirtenberger Patronenfabrik*, then one of the world's leading arms producers. Her life as a trophy wife was very unhappy. As Frau Mandl, Hedy was at the center of Viennese high society (at a time when the word "high" really meant it); but the trade-off was that her husband forbade her to pursue her acting career.

Allegedly, Herr Mandl was an obsessive control freak and jealous to the point of paranoia. He forced his young wife to be at his side, wherever he was, any time of the day and night. Thus Hedy had to sit long hours through her husband's business meetings. As the reader will soon see, she learned a lot.

She also became increasingly disturbed by the realization that the name of her husband's game was the illegal re-armament of Nazi Germany. The rest reads like a script for a bad Hollywood movie: Hedy drugs her maid and, in the maid's dress, escapes through a window. She goes to Hollywood, where she resumes her stardom, being at one point voted by the film critics the most beautiful woman in the world.

In 1940 (at the peak of America's isolationist stand in World War II), at a dinner party in Hollywood, Hedy met film-score composer George Antheil, and that was the point where the story dramatically deviated from the Hollywood stereotype. Hedy realized that George Antheil had crucial technical expertise: he wrote music for the player piano (electromechanical piano).[1]

You would not expect such behavior from a Hollywood star, but Hedy was thinking about an important technical problem: how can one safely control a torpedo by radio, without being jammed by the enemy? Her answer was *frequency hopping*: both the transmitter on the torpedo boat and the receiver in the torpedo should synchronously hop from frequency to frequency, so that the enemy is lost and does not know which frequency to jam.

George Antheil knew how to record music for an electromechanical piano using the arrangements of punched holes in a paper roll; instead of the piano, the punch roll readers could control a radio transmitter and receiver.

It took some time for Hedy and George to work out the details, but, by the height of the War, in 1942, they received a patent for their invention.

What happened next? In short, nothing. The invention was ignored (what else would you expect from Navy folks dealing with an actress and a piano player?) and was reinvented in the 1950s for missile guidance systems. However, nowadays Hedy is feted as the creator of modern communications technology. Mobile phones, indeed, use a much advanced version of her brilliant idea: rather than hopping, the signal is spread over several frequencies simultaneously.

8.2 Mathematics of frequency hopping

Frequency hopping immediately poses very serious mathematical questions.

- How can one make the sequence of hops unpredictable?
- What is the optimal jam-proof sequence of hops?
- How can one put as many phone calls in the given bandwidth as possible?

Thorough answers to these questions require highly sophisticated mathematical machinery. Meanwhile, some of the solutions (especially those developed at the very dawn of the new technology) are surprisingly simple and stunningly beautiful.

While Lamarr's patent was slowly expiring without use, frequency hopping technology was being independently developed by the Navy, for use in sonar systems. Rather than avoiding jamming, the purpose of frequency hopping in sonar location is to distinguish between reflected signals. Indeed, the time delay of the echo reflected from an underwater object allows us to measure the distance to the object. Due to the Doppler effect, sound waves reflected from a moving object change their frequency. Therefore sonar should, in principle, be able to tell both the distance to the target and its speed. It does so by sending a series of sounds of varying frequencies (to distinguish between echoes reflected from targets at different distances). The same sequence may be repeated many times.[2]

Figures 8.1–8.4 demonstrate some of the possible difficulties in reading echoes. It is not always possible to distinguish between a reflection from, say, a distant stationary object (Figure 8.2, shift in time, that is, along the horizontal axis) or a closer moving object (Figure 8.3, shift in frequency, that is, along the vertical axis). Look at Figure 8.4, where the two reflected signals are shown side by side.

Fig. 8.1. A primitive frequency-time pattern for hydroacoustic applications: at different time periods, the sonar emits signals of various frequencies. This makes it possible to distinguish between echoes reflected from targets at different distances.

However, if you use for your sonar the frequency-time pattern of Figure 8.5, you can easily see that it has the property that any two distinct shifts (vertical, or horizontal, or a combination of both) have at most one "ping" in common. This property allows one to distinguish between various kinds of reflection and to read echoes with ease. The frequency-time patterns with these properties are called *Costas arrays* [335, 356, 357, 358, 397].

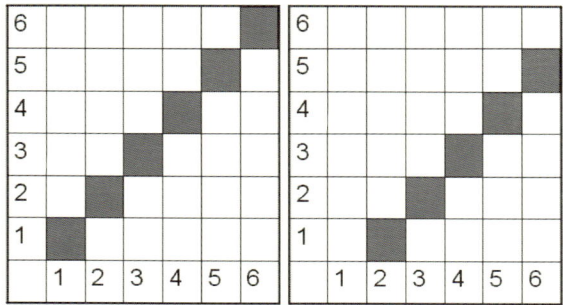

Fig. 8.2. Echoing from an object, the sequence returns shifted in *time*. The length of the time delay depends on the *distance* to the object.

Fig. 8.3. Echoing from a moving object, the sequence returns shifted in *frequency* which depends on the *speed* of the object.

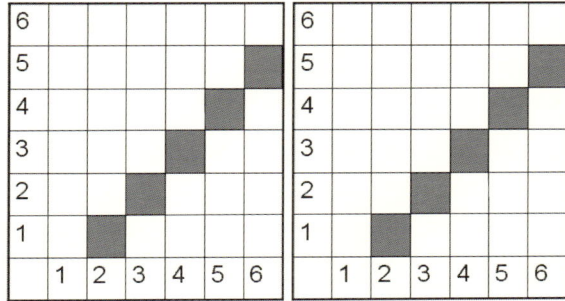

Fig. 8.4. Two echoes: a distant stationary object or a closer moving object?

8.3 "Kolmogorov's Logic" and heuristic reasoning

I wish to use the derivation of a special class of Costas arrays, Welch-Costas arrays (like the one in Figure 8.5), as an illustration of some useful principles of the heuristic argument.

Formal proofs in mathematics are based on the *modus ponens* rule (the Latin name goes back to medieval scholastic logic):

> if statement A implies statement B and if A is true, then B is also true,

or, in the now-standard notation of mathematical logic,

$$A \to B, A \vdash B.$$

The famous mathematician Andrei Kolmogorov made an unfortunate joke assigning the name *The "Woman's Logic" Principle* to the following widely used way of arguing:

Fig. 8.5. A Welch-Costas array: after any vertical (frequency) and/or horizontal (time) shift the new pattern and the old one have no more than 1 ping in common.

> *if statement A implies statement B and B is nice, then A is true.*

I leave it to the reader to decide whether women are more prone to committing this fallacy than men. Nevertheless, since the principle is of extreme importance for mathematical heuristics, it would be fair to call it *Kolmogorov's Principle*.

Indeed, if you are not looking for universal truth, but are concerned only with finding *one* solution of a concrete problem (and if it does not matter for you which particular solution you find— anything goes), the following version of Kolmogorov's Principle is quite useful:

> *if you want to achieve B and you know that A yields B, then try to achieve A first.*

Returning to Hedy Lamarr, we see that in the area of invention the principle becomes

> *if you want to make B and you know that the thing A, if it exists, would produce B, then try to make A first.*

It is hard to avoid the conclusion that this was Hedy Lamarr's way of thinking:

- B: frequency hopping masks the communication between the torpedo boat and torpedo, which is nice.
- A: the music of a player piano is nothing but hops from frequency to frequency.

- Hence try to squeeze the mechanism of a player piano into the torpedo.

This way of thinking is very efficient in solving mathematical problems, and I will try to illustrate it by looking at the way the pattern in Figure 8.5 was constructed. You will soon see that our preparedness to accept, for the sake of the desired conclusion B, an atrociously preposterous premise A may lead to very beautiful and unexpected solutions. *However, this approach does not guarantee that you will find **all** solutions. It does not ensure that you will find any solution at all. But it works—more frequently than not.*

Being mathematicians and knowing nothing else, we start by removing the real-world content from the problem and treating it as an abstract mathematical problem. To that end, we interpret the frequency-time pattern as the graph of a function, say,

The preparedness to accept, for the sake of the desired conclusion, an atrociously preposterous premise may lead to very beautiful and unexpected solutions.

$$y = f(x),$$

defined on a short initial segment of natural numbers, say,

$$X = \{\, 1, 2, \ldots, n \,\},$$

and taking only natural number values. In Figure 8.5, $n = 6$. As usual, the graph shifted b units to the right and a units in the vertical direction represents the function

$$y = a + f(x - b).$$

We want our pattern to have at most one "ping" in common with every non-trivial shift; to achieve that, it will suffice to have the property that every equation

$$f(x) = a + f(x - b) \tag{8.1}$$

where at least one of the parameters a, b is different from 0 has at most one solution.

What kind of equations definitely have at most one solution? Of course, linear equations,

$$Ax + B = 0.$$

Therefore the idea is to make our equation (8.1) be *equivalent* to a linear equation after some rearrangement.

How can we make the desired rearrangement? Notice that we still know nothing about the function $y = f(x)$, but we are prepared to make a list of the features which we would like it to have. Since we want to solve equations involving our (still unknown) function f, it will definitely be nice if $y = f(x)$ has an inverse function, that is, if there is a function $x = g(y)$ such that, for all $x \in X$,

$$x \equiv g(f(x)).$$

Then we can apply g to both sides of (8.1) and rewrite it as

$$x = g(a + f(x - b)). \qquad (8.2)$$

Well, it would be nice if this equation could be simplified and turned into a linear equation, something like

$$x = c(x + d).$$

Can we turn our equation into this form if we assume some nice properties of the functions g and f? We need an identity of the form

$$g(a + f(x - b)) \equiv c(x + d)$$

for some constant c and d computed from a and b. For that, we have to be able to place the function symbols g and f together:

$$g(\cdots + f(\cdots)) \equiv \cdots g(f(\cdots)).$$

We will be able to do that if $g(s + t) = g(s) \cdot g(t)$, because in that case

$$g(a + f(x - b)) = g(a) \cdot g(f(x - b))$$
$$= g(a) \cdot (x - b),$$

and (8.2) is reduced to the linear equation

$$x = g(a) \cdot (x - b) \qquad (8.3)$$

as desired. Hence we want the function g to satisfy the identity

$$g(s + t) \equiv g(s) \cdot g(t), \qquad (8.4)$$

that is, to be an exponential function: for if we denote $g(1) = d$, then (8.4) implies
$$g(x) = d^x \qquad (8.5)$$

for all $x \in X$. The problem is that the image of $X = \{1, \ldots, n\}$ under any ordinary exponential function almost definitely does not belong to X.

Can we still save the day? Yes; since we need only the identity

$$g(s + t) = g(s) \cdot g(t),$$

and we need only the values of g at small integers, let us make calculations in modular arithmetic, reducing every result modulo some fixed integer m:

$$g(x) = d^x \mod m.$$

For example, if we take $m = 7$ and $d = 3$ (in that case $n = 6$), we start with the equalities which define the function $g(x) = d^x$:

$$3^1 = 3 \qquad \mod 7,$$
$$3^2 = 9 = 2 \quad \mod 7,$$
$$3^3 = 27 = 6 \quad \mod 7,$$
$$3^4 = 81 = 4 \quad \mod 7,$$
$$3^5 = 5 \qquad \mod 7,$$
$$3^6 = 1 \qquad \mod 7.$$

The function $f(x)$, being the opposite of exponentiation, is the *discrete logarithm modulo* 7. Using the usual definition of the logarithm as the operation opposite to exponentiation, we have, for $f(x) = \log_3 x \mod 7$,

$$\log_3 3 = 1,$$
$$\log_3 2 = 2,$$
$$\log_3 6 = 3,$$
$$\log_3 4 = 4,$$
$$\log_3 5 = 5,$$
$$\log_3 1 = 6.$$

The frequency-time pattern in Figure 8.5 is exactly the graph of the discrete logarithm function $y = \log_3 x \mod 7$. $\boxed{[?]}$

In the notes, I rerun our heuristic argument in reverse order, making it into a rigorous proof.[3]

What we just did is a classical construction, known as a Welch-Costas array [335]. It works for every prime number p in place of the number 7 used in our example.

Indeed, it is a standard fact of elementary number theory that there exists a positive integer $d < p$ such that the modular exponential function $d^k \mod p$ takes every value in the range $1, 2, \ldots, p-1$, i.e.,

$$\{ d^1, d^2, \ldots, d^{p-1} \mod p \} = \{ 1, 2, \ldots, p-1 \},$$

and which therefore can be used as the basis of the discrete logarithm modulo p. Therefore a Welch-Costas array for p is the graph of the function

$$\{1, 2, \ldots, p-1\} \to \{1, 2, \ldots, p-1\}$$
$$x \mapsto \log_d(x) \mod p.$$

The following appears to be a hard open problem: does there exist a Costas array of size

$$32 \times 32?$$

See [357, 397] for more open questions.

Notice that, at every step of our argument, we were squeezing the player piano into the torpedo: we knew something from a completely different area of mathematics which appeared to resolve the immediate technical difficulty and were trying to emulate this something in our construction. And it worked and produced a non-trivial solution! [?]

> *Construct Welch-Costas arrays* mod 5, mod 11.

8.4 The triumph of the heuristic approach: Kolmogorov's "5/3" law

I wish to dispel the impression which I possibly created inadvertently that Andrei Kolmogorov was a die-hard formalist who did not value heuristic arguments. Quite the contrary; he was the author of what remains the most striking and beautiful example of a heuristic argument in mathematics. The deduction of his seminal "5/3" law for the energy distribution in the turbulent fluid [376] is so simple that it can be done in a few lines. Moreover, since it involves dimensional analysis, it is directly related to the arithmetic of named numbers; see Section 4.7. I was lucky to study at a good secondary school where my physics teacher (Anatoly Mikhailovich Trubachov, to whom I express my eternal gratitude) derived the "5/3" law in one of his improvised lectures. In my exposition, I borrow some details from Arnold [3] and Ball [257] (where I have also picked up the idea of using a woodcut by Katsushika Hokusai, Figure 8.6, as an illustration).

Fig. 8.6. Multiple scales in the motion of a fluid, from a woodcut by Katsushika Hokusai titled *The Great Wave off the Coast of Kanagawa* (from the series *Thirty-six Views of Mount Fuji*, 1823–29). This image is much beloved by chaos scientists. Source: *Wikipedia Commons*. Public domain.

The turbulent flow of a liquid consists of vortices; the flow in every vortex is made of smaller vortices, all the way down the scale to the point where the viscosity of the fluid turns the kinetic energy of motion into heat (Figure 8.6). If there is no influx of energy (like the wind whipping up a storm in Hokusai's woodcut), the energy of the motion will eventually dissipate and the water will stand still. So, assume that we have a balanced energy flow: the storm is already at full strength and stays that way. The motion of a liquid is made of waves of different lengths; Kolmogorov asked the question, what is the share of energy carried by waves of a particular length?

Sergey Utyuzhnikov, aged 12

Here is a somewhat simplified description of his analysis. We start by making a list of the quantities involved and their dimensions. First, we have the *energy flow* (let me recall, in our set-up it is the same as the dissipation of energy). The dimension of energy is

$$\frac{\text{mass} \cdot \text{length}^2}{\text{time}^2}$$

(remember the formula $K = mv^2/2$ for the kinetic energy of a moving material point). It will be convenient to make all calculations *per unit of mass*. Then the energy flow ϵ has dimension

$$\frac{\text{energy}}{\text{mass} \cdot \text{time}} = \frac{\text{length}^2}{\text{time}^3}.$$

For counting waves, it is convenient to use the *wave number*, that is, the number of waves fitting into the unit of length. Therefore the wave number k has dimension

$$\frac{1}{\text{length}}.$$

Finally, the *energy spectrum* $E = E(k)$ is the quantity such that, given the interval $\Delta k = k_1 - k_2$ between the two wave numbers, the energy (per unit of mass) carried by the waves in this interval should be approximately equal to $E(k_1)\Delta k$. Hence the dimension of E is

$$\frac{\text{energy}}{\text{mass} \cdot \text{wavenumber}} = \frac{\text{length}^3}{\text{time}^2}.$$

To make the next crucial calculations, Kolmogorov made the major assumption that amounted to saying:[4]

> The way bigger vortices are made from smaller ones is the same throughout the range of wave numbers, from the biggest vortices (say, like a cyclone covering the whole continent) to a smaller one (like a whirl of dust on a street corner).

Then we can assume that the energy spectrum E, the energy flow ϵ, and the wave number k are linked by an equation which does not involve anything else. Since the three quantities involved have completely different dimensions, we can combine them only by means of an equation of the form

$$E(k) \approx C\epsilon^x \cdot k^y.$$

Here C is a constant; since the equation should remain the same for small scale and global scale events, the shape of the equation should not depend on the choice of units of measurements; hence C should be dimensionless.

Let us now check how the equation looks in terms of dimensions:

$$\frac{\text{length}^3}{\text{time}^2} = \left(\frac{\text{length}^2}{\text{time}^3}\right)^x \cdot \left(\frac{1}{\text{length}}\right)^y.$$

After equating lengths with lengths and times with times, we have

$$\text{length}^3 = \text{length}^{2x} \cdot \text{length}^{-y},$$
$$\text{time}^2 = \text{time}^{3x},$$

which leads to a system of two simultaneous linear equations in x and y,

$$3 = 2x - y,$$
$$2 = 3x.$$

This can be solved with ease and gives us

Alexander Chorin
(the youngest child),
aged 3

$$x = \frac{2}{3} \quad \text{and} \quad y = -\frac{5}{3}.$$

Therefore we come to *Kolmogorov's "5/3" law*:

$$E(k) \approx C\epsilon^{2/3}k^{-5/3}.$$

The dimensionless constant C can be determined from experiments and happens to be pretty close to 1.[5]

The status of this celebrated result is quite remarkable. In the words of an expert on turbulence, Alexander Chorin [331],

> Nothing illustrates better the way in which turbulence is suspended between ignorance and light than the Kolmogorov theory of turbulence, which is both the cornerstone of what we know and a mystery that has not been fathomed.
>
> The same spectrum [...] appears in the sun, in the oceans, and in manmade machinery. The 5/3 law is well verified experimentally and, by suggesting that not all scales must be computed anew in each problem, opens the door to practical modelling.

Arnold [3] reminds us that the main premises of Kolmogorov's argument remain unproven—after more than 60 years! Even worse, Chorin points to the rather disturbing fact that

> Kolmogorov's spectrum often appears in problems where his assumptions clearly fail. [...] The 5/3 law can now be derived in many ways, often under assumptions that are antithetical to Kolmogorov's. Turbulence theory finds itself in the odd situation of having to build on its main result while still struggling to understand it.

8.5 Morals drawn from the three stories

Shall we draw some conclusions from the three stories told in this chapter?

In the case of Welch-Costas arrays, our assumptions were rather timid, and afterwards we had no difficulty in converting the heuristic argument into a rigorous proof that the proposed solution was, indeed, a solution. There is nothing surprising in this since we were dealing with a problem which was at the opposite end of the mathematical spectrum from the great enigma of turbulence.

In Kolmogorov's case, the boldness and the dramatic scale of his heuristic assumptions led to equally dramatic and paradoxical results, and mathematicians still struggle to comprehend why his argument works.

The Hedy Lamarr story tells us that an outrageous idea, even if it does not lead to a practically workable solution, may show the direction for future developments.

In all three cases we see that in mathematics (and in the general area of invention) it pays to be bold.

8.6 Women in mathematics

Since I have quoted Kolmogorov's disparaging remark on women's reasoning, I feel that I have to clarify my position on the thorny issue of the role and place of women in mathematics.

Mathematics is a weapon of personal empowerment.

It is a sad statistical fact that the number of women doing mathematical research is disappointingly small. I am confident that this cannot be explained by the psychophysiological gender differences—even if neurophysiologists find subtle variance in language and visual processing in male and female brains [154, 199].[6]

The problem, I believe, is sociopsychological rather than psychophysiological. I'll try to outline, briefly, my vision of it.

A rarely discussed side effect of doing mathematics is that mathematics is a *weapon of personal empowerment*. To be successful in mathematics, you have to be bold, you have to be absolutely independent in your thinking. When you prove something new, you are in the unique position of being the only person on the Earth who knows the Truth— and is prepared to defend it. On the other hand, the principles of mathematical rigor give you the right to question whatever other mathematicians say. If this still does not sound to you as a recipe for trouble, you can also take into consideration that research mathematics is fiercely competitive. This is an explosive brew.

> *Mathematicians tend to forget how psychologically tense and charged mathematical discourse is.*

Mathematicians who grew up in this chivalrous environment tend to forget (or ignore) how psychologically tense and charged mathematical discourse is. This becomes apparent only in comparison with other walks of life.

> *To recognize someone as a fellow mathematician means to accept that she is intellectually equal (or even superior) to you. Too many men will still feel uncomfortable with that.*

The example I want to give is probably extreme. Once I stayed with a colleague at Princeton over a long and lazy Labor Day weekend. My hosts and I were invited to their neighbor's garden party, where I found myself in the company of twelve professional astrologists, exquisitely groomed ladies with loads of heavy silver jewelry, mostly Zodiac signs. Besides the usual party talk, the astrologists actively discussed, between themselves, matters of their professional interest. It was fun to watch; their chat, saturated by astrological jargon, sounded, to a lay observer like me, almost like a chat between mathematicians, but with a surreal feel to it. With some effort, I finally realized what made it surreal to me: *they immediately believed and accepted everything that their colleagues were saying to them*; on their faces, there were none of mathematicians' usual expressions of tightly focused mistrust. I realized why my daughter had told me that she was scared to be present during my professional conversations with mathematician friends: in her words, we looked as if we were ready to fight each other.

Mathematics is highly psychologically charged and competitive, but fights remain invisible for the onlooker and are strictly ritualized by a very strong research ethic and the principles of mathematical rigor. Arguments are rarely linked to money and, therefore, do not lead to serious bloodletting. Mathematicians usually look in disgust at the morals in many other, more practical, disciplines, where the high cost of research (and the scarcity of funding) and the lack of clear criteria of rigor naturally instill a dog-eat-dog mentality.

When money gets involved, everything becomes depersonalized: what matters is not who you are but what is your place in the pecking order. We are all accustomed to seeing fools in high places, and although the spectacle is rarely pleasant, it does not get deep under the skin. A male chauvinist can tolerate a woman in a position of superiority by treating her as yet another case of undeserved promotion.

The crucial difference of mathematics from many other walks of life is that its power games are deeply personal in the purest possible sense. To recognize someone as a fellow mathematician means to accept that she is intellectually equal (or even superior) to you and that she has the right to wear, like a knight's armor, her aura of intellectual confi-

In British schools, many teachers of mathematics routinely suppress mathematically able students because the students' intellectual superiority makes the teachers feel insecure.

dence and independence. Too many men still feel uncomfortable with that.

Unfortunately, even gender study researchers are sometimes uncomfortable with the principle of intellectual independence when mathematics is concerned. I was surprised to read, for example, the following:

> The classroom structure, designed to foster independent non-collaborative thinking, is most supportive of white male, middle-class socialization models, and it continues through university (Pearson & West, 1991). It encourages sex-role stereotyped forms of communication—independence, dominance, assumption of leadership—in which males have been trained to excel. Women, conversely, feel uncomfortable and excluded in situations requiring such behavior; yet, their participation—as questioners as well as newly-minted authorities—may be critical to knowledge acquisition and school success. The importance that women place on mutual support, building collaborative knowledge, and applying it practically is devalued in comparison with the importance of individual expertise to males and their inclination to debate abstract concepts. [452]

Of course, mathematics is all about "independent non-collaborative thinking". But why should we assume that women are less capable of independent thinking? Why should women surrender the game without a fight?

Women remain a disadvantaged group of our society; but when you look at the even more disadvantaged and vulnerable group of *children*, you find something even more striking.

I was shocked to hear from several leading British experts on mathematical education that, in British schools, many (if not most) teachers of mathematics routinely suppress mathematically able students because the students' intellectual superiority makes the teachers feel insecure.

Inna Korchagina,
aged 6

The complex of "mathematical insecurity" in secondary school teachers of mathematics is lamentable, but not unexpected. But I was quite surprised to hear from my mathematician colleagues that mathematical insecurity is a factor of academic politics at the university and national levels: it explains—my colleagues claimed to me—a surprising level of hostility to pure mathematics found in some senior representatives of the engineering mathematics community.

I would bet that the suppression of able children happens in most other school subjects, but mathematics sets up the scene where it could be obvious for both the teacher and the student that the student is superior; British teachers are not trained to handle such conflicts with dignity and respect to the child. I shall be forever grateful to my mathematics teacher in my village school in Siberia, Aleksandra Fedotovna Lazutkina, who frankly told me (in front of the class!) that she could not teach me anything and asked me to somehow teach myself. After that, I sat quietly in her classes, minding my own business; she had never asked me a single question, or otherwise interfered with my work. I cannot imagine a British or American teacher of today behaving the same way.

> We still live in a culture where women are disapproved and penalized if they show real *intellectual independence*.

Unfortunately, we still live in a culture where women are allowed to play, on an equal footing with men, the conformity games in the office or even in politics but are disapproved and penalized if they show *real* intellectual independence.

It was even more true in Hedy Lamarr's times. The fact that she was not a mathematician only supports my main point: indeed, what mattered is that she exhibited an outrageous intellectual independence. For a woman in America, in 1940, to go against the political mainstream and seriously think about preparation for the looming war with Germany was strange, to put it mildly; to think seriously about radio controlled torpedoes—that was just insane. By 1942, at the height of the war, the military, political, and domestic economic situations changed. American munition factories desperately needed female workers to replace conscripted men, and Hedy was glorified by the official propaganda machine. Being propaganda fodder though did not help her invention to be taken seriously by the Navy.

I do not know an easy way to change the position of women in mathematics. I would suggest, tentatively, that when promoting mathematics, we should put more stress on its personal empowerment aspect; we should encourage competitiveness and independent thinking; we should openly talk to our students about the power games of mathematics. It does not fit easily into the existing policy of mathematical education, but it is worth trying.

Maria Leonor
Moreira,
aged 11

Notes

[1]PLAYER PIANOS AND OTHER ELECTRICAL DEVICES. As a side remark, it is worth mentioning that electrical devices like player pianos allow musicians to design and use more flexible scales; in this context, the subtle mathematical nature of musical scales becomes really essential. See the charming paper by Wilfrid Hodges [51] for more detail. Apparently, George Antheil was seriously intrigued by the aesthetics of the new age of machines; in his earlier years, he composed the score for the famous cubist film *Ballet Mecanique* (the film was directed by Fernand Léger, one of the leaders of the cubist movement).

[2]HYDROACOUSTICS AND WHALE BEACHING. It is now widely accepted that unfortunately the main reason for whales beaching (when whales throw themselves on the shore) is that the poor whales are driven to mass suicide by the intolerable level of noise created by the sonar and hydroacoustic communication systems of submarines. Submarines use the same bandwidths that whales have used for millennia for their own communication.

[3]CONSTRUCTION OF WELCH-COSTAS ARRAYS. First of all, we have to clarify the nature of the our logarithm. We use the Fermat Theorem:

$$x^{p-1} \equiv 1 \pmod{p}.$$

Hence $(x, y) \mapsto x^y$ is well-defined as a function from

$$(\mathbb{Z}/p\mathbb{Z} \smallsetminus \{0\}) \times (\mathbb{Z}/(p-1)\mathbb{Z})$$

into

$$\mathbb{Z}/p\mathbb{Z} \smallsetminus \{0\}.$$

But $\mathbb{Z}/p\mathbb{Z} \smallsetminus \{0\}$ is the multiplicative group of the field $\mathbb{F}_p = \mathbb{Z}/p\mathbb{Z}$, and if a is in this group, then

$$y \mapsto a^y$$

is a homomorphism from $\mathbb{Z}/(p-1)\mathbb{Z}$ into $\mathbb{F}_p{}^\times$. We can pick a as a *generator*, and then the homomorphism is an isomorphism, so it has an inverse, $\log_a$.

This means that, in our example, the logarithm takes values in the additive group of residues modulo 6; it will be prudent to use a different symbol, "$\oplus$", for addition modulo 6, to distinguish it from the addition "$+$" modulo 7. Notice that, in our graphs, the horizontal and the vertical axes represent different algebraic structures! This happens because the logarithm is an isomorphism of the *multiplicative* group of residues modulo 7 onto the *additive* group of residues modulo $7 - 1 = 6$.

However, for our toy logarithm, we still have the usual identities

$$3^{\log_3 x} = x \quad \text{mod } 6, \qquad \log_3(ab) = \log_3 a \oplus \log_3 b \quad \text{mod } 6.$$

Moving further, how many points do the graph of the function

$$y = \log_3(x)$$

(our pattern) and the pattern formed by reflection from a moving target have in common? We already know that the reflection corresponds to the graph of the function

$$y = a \oplus \log_3(x + b).$$

The common points of the two graphs correspond to the solution of the equation

$$\log_3 x = a \oplus \log_3(x + b) \quad \text{mod } 6.$$

If we now do what we do with ordinary logarithmic equations of precalculus and exponentiate both parts of the equation,

$$3^{\log_3 x} = 3^{a \oplus \log_3(x+b)} \quad \text{mod } 7,$$

we come to

$$x = 3^a(x + b) \quad \text{mod } 7,$$

a linear equation in x, which has a unique solution unless $3^a = 1$ and $b = 0$, which means that $a = 0 \mod 6$ and $b = 0 \mod 7$, and hence there was no shift in the first place.

[4]This formulation is a bit cruder than most experts would accept; I borrow it from Arnold [3].

[5]HISTORY OF DIMENSIONAL ANALYSIS. It would be interesting to have an account of the history of dimensional analysis. It can be traced back at least to *Froude's Law of Steamship Comparisons*, used to great effect in D'Arcy Thompson's book *On Growth and Form* [93, p. 24] for the analysis of speeds of animals: the maximal speed of similarly designed steamships is proportional to the square root of their length. William Froude (1810–1879) was the first to formulate reliable laws for the resistance that water offers to ships and for predicting their stability.

[6]GENDER DIFFERENCES. See the transcript of an illuminating debate between Steven Pinker and Elizabeth Spelke, http://www.edge.org/3rd_culture/debate05/debate05_index.html.

9

Recovery vs. Discovery

For in much wisdom [is] much grief:
and he that increaseth knowledge
increaseth sorrow.
Ecclesiastes 1:18

This chapter contains some technical parts that can be safely skipped in the first reading.

9.1 Memorize or rederive?

Mathematics provides very efficient methods of recovering mathematical facts, much more efficient than straightforward memorization. I remember the *result* of hardly any arithmetic calculation that I have done in my life, but if I have to do one of them again, I shall get the same result. Many of my colleagues are prepared to admit that they do not remember any trigonometric formulae beyond the most basic ones, but, if necessary, they can recover and prove most formulae of elementary trigonometry, with relative ease.

This aspect of mathematical practice is sometimes completely lost on our students and is not so frequently discussed in the professional literature. The following quotation is taken from a real student examination script (UMIST, January 2004). Asked to prove that $\sqrt{2}$ is irrational, the student responded:

> The only proof I can offer is that I remember that $\sqrt{2}$ is irrational, according to *1984*, human memory is more important than proof as proof can be altered, memory cannot.

Compare this with another extreme opinion, this time of a professional mathematician (Kevin Coombes [14]):

Never memorize a formula if you can find a way to rederive
it.[1]

I argue that there is a difference between the *discovery* of new
mathematical facts and *recovery* of forgotten ones.

Reuben Hersh, in his talk at the *Philosophy of Mathematical
Practice* conference in Brussels in September 2002 [50], showed a
nice example of what I call *recovery technique*: the (re)derivation of
Heron's formula for the area of the triangle[2]. Writing the present
book more than two years later, I decided to use Hersh's method
to recover the formula while honestly recording my stream of consciousness. The experiment, I hope, was sufficiently clean: I had
known Heron's formula when I was at school but had not used it
for ages and remembered it only in the vaguest terms.

Reuben Hersh's talk was aimed
at philosophers of mathematics and
purported to give an example of a creative process and discovery in mathematics. Since I was not a philosopher
and had heard about Heron's formula
before, I saw in the procedure something different, not discovery but recovery. Indeed, I believe that recov-

> I believe that recovery is a highly special-
> ized activity; the discovery of new math-
> ematics is done differently. You know, the
> feel is very different.

ery is a highly specialized activity; the *discovery* of *new* mathematics is done differently. You know, the feel is very different.

There are, however, deeper reasons to expect recovery to be different from discovery. In the case of recovery, we, as a rule, already
know which mathematical language we used for the formulation
of the result; in the case of discovery of non-trivial mathematical
facts, the search for an appropriate language frequently happens
to be the most challenging part of the job. See Section 4.2 for a
discussion of cryptomorphisms and multiplicity of languages, and
Chapter 10 for more examples of problems whose solutions involve
"search for language".

It is time to fix some terminology. A *recovery procedure* is a set
of heuristic rules which we vaguely remember to apply when we
want to recover a mathematical fact. It is like a poster on the control panel of some serious machine, a submarine or a plane, which
says what the crew should do when things have gone haywire. A
rederivation is a semi-heuristic argument made in accordance with
the recovery procedure.

For example, when teaching calculus, I insist that my students
help me to recall standard trigonometric formulae; I honestly admit
that I remember hardly any beyond the most fundamental one,

$$\sin^2 \alpha + \cos^2 \alpha = 1.$$

However, I (like most professional mathematics teachers) can derive more or less every standard formula. Also, I can almost immediately tell when the formula given to me by a student is wrong. It is quite safe to play this game with students: ask them to give *wrong* formulae and then explain to them immediately, on the spot, why the formulae are wrong. Indeed, students are hopeless at the production of plausible formulae; if a student's formula is wrong, it is demonstrably, spectacularly wrong. I have several recovery procedures for use in trigonometry: nothing special, really; the one for the formulae of the type $\sin(\alpha + \beta)$ consists of multiplying two rotation matrices—or complex multiplication. The formula for the matrix product is more fundamental than any trigonometric formula—with the possible exception of

$$\sin^2 x + \cos^2 x = 1;$$

see the note *Simplest possible examples* on page 19.

9.2 Heron's formula

Now I am ready to outline a recovery procedure for Heron's formula. Suppose you want to reproduce the formula for the area S of a triangle in terms of the lengths x, y, z of its sides. You vaguely remember that the formula is something like

$$S = \sqrt{(\text{a kind of polynomial in } x, y, z)}.$$

Your recovery procedure consists of two general principles:

- use the symmetry properties of the polynomial resulting from the symmetries of the problem; and
- look at the degenerate cases when the triangle collapses into a segment and its area vanishes.

I can now describe the rederivation based on these rules. Denoting the polynomial under the root by $F(x, y, z)$ and squaring both sides, you come to

$$S^2 = F(x, y, z)$$

and need to find the polynomial $F(x, y, z)$. You know, of course, that the triangle degenerates into a segment when one of the sides x, y, z equals the sum of other two, say, $z = x + y$; hence $F(x, y, z) = 0$ if $x + y - z = 0$. You conclude that it is likely that $F(x, y, z)$ is divisible by $x + y - z$, etc., and that therefore

$$F(x, y, z) = (x + y - z)(x - y + z)(-x + y + z)G(x, y, z)$$

for some other polynomial $G(x, y, z)$. Also, the dimensional considerations tell us that S^2 should be of degree 4; hence $G(x, y, z)$ is a linear function in x, y, z. Since x, y, z should appear in a completely symmetrical fashion, this means that

$$G(x, y, z) = a(x + y + z)$$

for some coefficient a and

$$F(x, y, z) = a(x + y - z)(x - y + z)(-x + y + z)(x + y + z).$$

For the equilateral triangle with sides $x = y = z$, the area is

$$S = \frac{\sqrt{3}}{4} x^2;$$

hence $a = 1/16$ and

$$S = \frac{1}{4}\sqrt{(x + y - z)(x - y + z)(-x + y + z)(x + y + z)}$$
$$= \sqrt{p(p - x)(p - y)(p - z)}$$

(at this point you are likely to recall the traditional form of Heron's formula, with

$$p = \frac{x + y + z}{2}$$

denoting the half perimeter of the triangle).

The choice of every step in this rederivation is made easier by vague memories of what the formula should look like. To make it into a rigorous proof, however, one needs a modicum of classical polynomial algebra; see Daniel Klain [285] for details. Actually, we are entering the realm of real algebraic geometry, since we have to judge the degree of "degeneration" of various special cases of the problem. Indeed, we need some reasons to conclude that the argument along the lines of

> Since $S^2 = F(x, y, z)$ vanishes when one of the sides x, y, z equals 0, we conclude that $F(x, y, z)$ is divisible by x, y, z and hence has the form $F(x, y, z) = xyz \cdot G(x, y, z) \ldots$

—very similar to the one we have just made—is invalid because the case $x = 0$ is excessively degenerate. We return to that point soon, in Section 9.4.

9.3 Limitations of recovery procedures

Unlike the discovery and proof of a seriously new result, a recovery procedure or a rederivation usually exists in an established conceptual framework. The validity of the result is known, and the issues of rigor, etc., are not that essential. Notice, in passing, that

the aesthetic status of recovery procedures is different from that
of proofs; we can tolerate an awkward and tortuous proof, while
recovery procedures, by their very nature, should be slick. Not surprisingly, many of them, once discovered, have been converted into
proper proofs. On the other hand, in many cases the recovery procedure is nothing but a specialization of a much more general but
better remembered result: it is a deduction rather than induction.

I quote a letter from Reuben Hersh, who uses my previously
mentioned example of recovery of the formula $\cos(\alpha+\beta)$ to comment
on the difference between recovery and discovery:

> I am working with a young friend, a high-school student
> here.
>
> Last week we needed the addition formulas for $\sin$ and
> $\cos$, in order to prove that angles add when you multiply
> complex numbers.
>
> Here's what I did.
>
> First, show geometrically, visually, that rotation is addi-
> tive (rotating the sum of two vectors is the same as adding
> the rotated vectors, it's just moving a rectangle as a rigid
> body). And, stretching or reflecting before or after rotating
> gives the same result. (No need to introduce the general
> concept of "linearity".) Then visually show that rotating the
> horizontal unit vector through α degrees produces the vec-
> tor $(\cos\alpha, \sin\alpha)$. Then the same for the vertical unit vec-
> tor, producing the vector $(-\sin\alpha, \cos\alpha)$. (For re-deriving, it's
> sufficient to choose α as an acute angle.)
>
> Finally, to rotate through $\alpha + \beta$, write an arbitrary unit
> vector as $\cos\beta$ times the unit horizontal vector plus $\sin\beta$
> times the unit vertical vector. Then separately rotate the
> horizontal and vertical components, and add. The result
> is a vector whose component in the horizontal direction is
> $\cos(\alpha + \beta)$, and component in the vertical is $\sin(\alpha + \beta)$.
>
> This does not require knowing in advance a represen-
> tation of the rotation operator. Of course, from a "higher"
> (operator-theoretical) point of view, it is just deriving such a
> representation. But as a way of deriving two trigonometric
> formulas, it has no prerequisites except the definition of the
> sine and cosine functions, the use of Cartesian coordinates,
> and adding vectors in the plane.

Of course, Hersh is absolutely right, but the whole point of my
recovery procedure is that I already *know* that the group of rotations can be represented by matrices and I already *know* the
principles of linearity. Ask any mathematician, what is he more
likely to forget: the addition formula for cosine or the definition of
a linear operator? The recovery procedures, like memory itself, are

something very personal, and I would not always recommend to my students the same procedures that I use myself. The hierarchy of mathematical principles and concepts as used by someone who has learned some mathematics is quite different from the one used by a novice of the same mathematical theory or discipline.

And a final comment on memory: motor skills, once acquired, are forever. It is virtually impossible to unlearn how to swim or ride a bicycle. I just wonder to what extent the same is true with respect to mathematical facts and concepts interiorized by engaging several different cognitive systems; do we remember them bet-

> *Ask any mathematician, what is he more likely to forget: the addition formula for cosine or the definition of a linear operator?*

ter? The attentive reader certainly noticed that Reuben Hersh's proof of the addition formula is a verbal description of a classical diagrammatic proof. It is an example of what I call, without assigning any derogative meaning, *proof by handwaving*: a proof which can be given orally, with the assistance of a few gestures, to help the visualization. For the listener, the proof might still be hard to comprehend; for the expositor, it will be welded, hardwired into his or her brain.

See more on handwaving and its elimination in Section 6.3.

9.4 Metatheory

Very often we have in our possession a metatheory which suggests and explains the structure of the desired fact. In the case of Heron's formula this metatheory is classical invariant theory: the problem has a natural symmetry, i.e., the sides x, y, z can be permuted in an arbitrary way under the action of the symmetric group Sym_3. The polynomial $F(x, y, z)$ is invariant under this action (that is, it does not change when we permute x, y, z), and so it belongs to the ring of polynomial invariants of Sym_3 in its natural action on the ring $\mathbb{R}[x, y, z]$ of polynomials in variables x, y, z. It is a classical result of algebra that the ring of polynomial invariants of this action is freely generated by the symmetric polynomials

$$1, \quad u = x + y + z, \quad v = xy + xz + yz, \quad w = xyz$$

(that is, u, v, w are not bound by any algebraic relation). This explains the ease with which we manipulated and factored the polynomials. Indeed, write

$$F(x, y, z) = G(u, v, w).$$

The key step of the recovery argument is the observation that if $G(u, v, w)$ takes value 0 at points (u, v, w) where

$$(-x + y + z)(x - y + z)(x + y - z) = -u^3 + 4uv - 8w = 0,$$

then $G(u, v, w)$ should be expected to be divisible by $-u^3 + 4uv - 8w$. (But we have to remember that this is still guesswork, not a proof, since u, v, w cannot take arbitrary values: recall, for example, that x, y, z are non-negative and hence $w \geqslant 0$. We will return to this point soon.)

There is a three-dimensional analogue of Heron's formula, the tetrahedron formula. Piero della Francesca was apparently the first mathematician to express the volume of the tetrahedron in terms of the edges [75]. Euler published the tetrahedron formula in 1758.[3]

It expresses the volume of a tetrahedron (simplex) in terms of the lengths of its six edges x, y, z, X, Y, Z (arranged in pairs of opposite edges (x, X), (y, Y), (z, Z)). In modern notation [306] it looks like

$$V = \begin{vmatrix} 0 & x^2 & y^2 & z^2 & 1 \\ x^2 & 0 & Z^2 & Y^2 & 1 \\ y^2 & Z^2 & 0 & X^2 & 1 \\ z^2 & Y^2 & X^2 & 0 & 1 \\ 1 & 1 & 1 & 1 & 0 \end{vmatrix}.$$

Unfortunately, the polynomial on the right-hand side has no non-trivial polynomial factorizations, so the recovery procedure as described for Heron's formula would not work here. Well, actually we do not need one—we hardly ever use Heron's formula, but who needs the tetrahedron formula in everyday mathematical life?

John Stillwell, aged 6

Invariant theory to some degree explains why we should not expect Piero della Francesca's formula to be easily recoverable: here, the group of symmetries of the problem is the symmetric group Sym_4, which acts by arbitrary permutations of the four vertices of the tetrahedron and causes the relabelling of the edges. However, the ring of invariants of the resulting action of Sym_4 on the polynomial ring $\mathbb{R}[x, y, z, X, Y, Z]$ is no longer free. Moreover, it is not a unique factorization domain—this immediately follows from [311, Theorem 3.9.2]. Another contributing factor to the failure of the recovery procedure in the three-dimensional case is the much more complicated geometry of the set of all possible tetrahedra, whereas in the two-dimensional case it is as simple as possible: just a simplicial cone C given by inequalities

$$0 \leqslant x \leqslant y + z, \qquad 0 \leqslant y \leqslant x + z, \qquad 0 \leqslant z \leqslant x + y.$$

The geometry of the cone C finally explains why our recovery procedure works but the argument at the end of Section 9.2 fails. Indeed, the intersection of the surface

$$-u^3 + 4uv - 8w = 0$$

with the cone C is the union of its faces

$$x = y + z, \qquad y = z + x, \qquad z = x + y.$$

On the other hand, the intersection of the surface $xyz = 0$ with the cone C consists only of the edges

$$0 = x \leqslant y + z, \qquad 0 = y \leqslant z + x, \qquad 0 = z \leqslant x + y$$

and does not allow us to conclude that $G(u, v, w)$ is divisible by $w = xyz$.

But this is not yet the end of the story: invariant theory also suggests that when we are dealing with *isosceles tetrahedra*, that is, tetrahedra with equal opposite edges $x = X$, $y = Y$, $z = Z$, we are back to the comfortable setting of Sym_3 acting on $\mathbb{R}[x, y, z]$. We can conclude therefore that the corresponding formula should be easily recoverable by essentially the same method as Heron's formula. This is done by Klain in [285, Theorem 2]; I do not quote the formula here and leave its recovery to the reader—as an exercise. I can however give a hint: when considering degenerate tetrahedra, go for the least dramatic. Do not collapse the tetrahedron into a segment, but just flatten it into a planar quadrangle (compare this with the two-dimensional case, where degeneration $x = 0$ was "less general" than $x + y = z$). [?]

> *Recover the formula for the volume of an isosceles tetrahedron.*

Finally, you may wish to use exactly the same method to try to guess the classical formula for the area of a quadrangle with sides a, b, c, d if it is known that it can be inscribed into a circle (*Brahmagupta's formula*). [?] Hint: the formula should be symmetric in terms of a, b, c, d—but why? Here is another useful observation: every triangle can be inscribed into a circle; therefore one should expect that Heron's formula is a special case $(d = 0)$ of Brahmagupta's formula.

> *Recover Brahmagupta's formula.*

As we see, our recovery procedure looks more and more like a little theory. Well, this is what we expect to happen with good recovery procedures—they eventually grow up into proofs and theories.

Notes

[1]Incidentally, O'Connor and Hermelin [214] analyzed the response time of idiot savant calendrical calculators on a variety of dates in the past

and the future; their interpretation of the results is that memory alone is inadequate to explain the calendrical calculating performance of the idiot savant subjects. It appears that a version of Coombes's principle works even at that level.

[2]Reuben Hersh's exposition [50] was fairly close to Alperin [252] and Klain [285]; it is worth noticing that the same argument has been independently discovered and documented at least three times.

[3]John Stillwell pointed out that [264, p. 285] wrongly states the date of publication as 1753; see paper E231 at `http://www.math.dartmouth.edu/~euler/`.

10

The Line of Sight

In this chapter, I present a case study of the life cycle of a mathematical problem—from its incidental birth through various reformulations coming from completely independent sources to absolutely unexpected applications. I have chosen this particular story simply because I can produce a first-hand account. The events described were driven by sheer human curiosity and appreciation of the problem's beauty. The problem is very elementary and belongs to the (mostly) unwritten tradition of "mathematical folklore" that lives outside the institutionalized framework of academic journals, conferences, etc. Indeed, it is precisely the elementary nature of the problem that allows us to see more clearly the mechanisms which drive the development of much more serious mathematical theories.

The key mathematical concept involved in my story is that of *convexity*. We shall see how prominent the semi-informal notion of the *line of sight* is in our thinking about convex bodies. In that sense, the chapter continues the discussion of lines of sight and convexity started in Section 4.4.

An interesting observation (perhaps relevant to the sociology of mathematics) is that the problem and the solutions were independently discovered by many people; as I learned many years after my first exploration, I was not the first one. There were much earlier contributors, for example Fejes Toth and Heppes [350], Danzer, and Dawson [262]. But I describe the story as it happened to me.

10.1 The Post Office Conjecture

The problem that I want to discuss in this chapter originates in 1977. At that time I was a student at Novosibirsk State University. My валенки (traditional Russian felt boots[1], a bit old-fashioned but still indispensable in the Siberian winter) developed a hole; the

197

only way of repairing them was to send them by parcel post back to my home village where the old cobbler still practiced the ancient art of felt boot patching. I packed the boots and took the parcel to the post office, where it was promptly rejected for reasons which can be formulated mathematically: the parcel was not convex. I struggled through the blizzard back to my dormitory and thought about the possible rationale behind the refusal to accept my parcel. This is how I came to my first conjecture.

Try to prove or disprove it without reading the succeeding text.

The Post Office Conjecture. A heap of (finitely many) convex parcels can be taken apart by removing one parcel at a time, without disturbing the rest of the heap. [?]

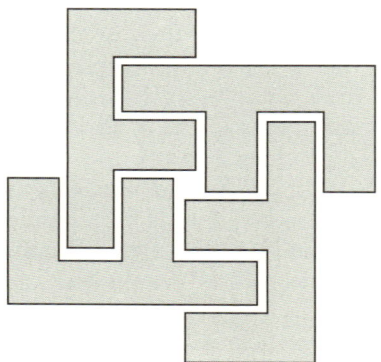

Fig. 10.1. For the Post Office Conjecture. This heap of non-convex letters F can be taken apart by slightly moving one letter, then another, then the first one again, etc., but touching only one letter at a time. However, the heap cannot be taken apart by pulling one letter at a time and in a single uninterrupted movement. Moreover, if the letters were a bit fatter and fit one into another perfectly, the first ("shake and take") method would not work, either; see Figure 10.2. Illustration by Ali Nesin.

In any case, it is fairly obvious that a heap of non-convex bodies cannot always be taken apart; if, in addition, we are allowed to pull only one body at a time, success is even less likely (Figure 10.1).

On my return to the dormitory, I shared my grief with my roommates, Eugene Khukhro and Serguei Karakozov. While I was repacking my parcel, they, helpful as always, tried to prove or refute the Post Office Conjecture—and failed. However, Karakozov, who was studying functional analysis, quickly proved the two-dimensional case. The next day he sought help from his senior colleagues at the Institute of Mathematics. After some discussion, the experts came to the conclusion that the conjecture was highly un-

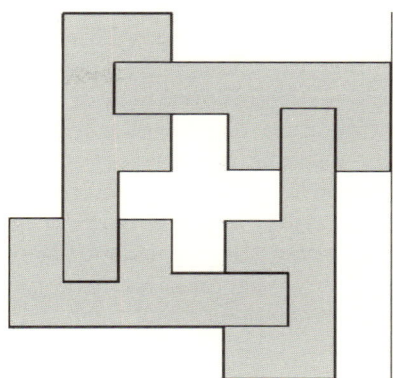

Fig. 10.2. For the Post Office Conjecture: "Shake and take" does not work. Illustration by Ali Nesin.

likely to be true in dimensions higher than 2, but failed to produce a counterexample.

However, the proof in the two-dimensional case is very simple; we discuss it a bit later.

Eugene Khukhro could not allow the problem to be wasted and made it into олимпиадную задачу, a mathematics competition problem. Here it goes:

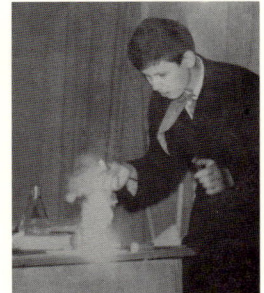

Eugene Khukhro, aged 14

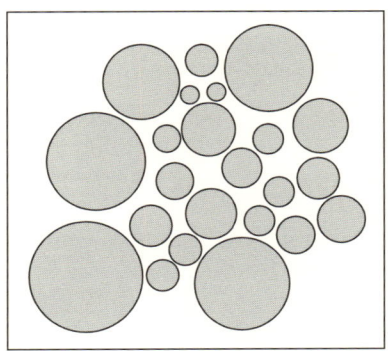

Fig. 10.3. For the Coins Problem. Illustration by Ali Nesin.

The Coins Problem. There are a finite number of round coins on the table, possibly of different sizes. The coins do not touch each other. Prove that it is possible to slide the coins along the surface of the table to its edge, one by one, without clicking a coin against another. (Figure 10.3)

What strikes me today is how obviously our way of life was imprinted onto this formulation. Around the clock, someone was sleeping in our room. Khukhro was a man of firm customs and slept from midnight to noon. I was moonlighting a bit as a night warden at the preparatory boarding school of our university (Физматшкола, a remarkable establishment; I mention it in this book on a few occasions). I returned from my duties at 9 in the morning and took a power nap until about two in the afternoon. However, Karakozov had a peculiar habit of going to bed 15 minutes later than on the previous day and lived in a continuous phase shift; eventually he reached the point where he went to bed when I got up. We behaved in a very considerate way: we never switched on the light in our room and even plugged an electric kettle in the corridor, to minimize the impact of its hissing and puffing on our sleeping friends.

But let us return to the problem. It was used in a mathematics olympiad in Siberia in about 1979–80, and some pupils solved it. The reader, probably, expects an elementary solution. What follows is one of many possible. Hammer a nail into the center of each coin, so that it sticks out but holds the coin in its place. Now imagine an elastic band stretched to form a large circle enclosing all the coins. As the band is allowed to contract, it will eventually come to rest in contact with some "extreme" nails (Figure 10.4). The "extreme" coins are exactly those which can be removed without clicking (prove it!). Then, of course, we can repeat the procedure until no coin is left on the table. As you can see, everything is quite straightforward. However, three remarks are due:

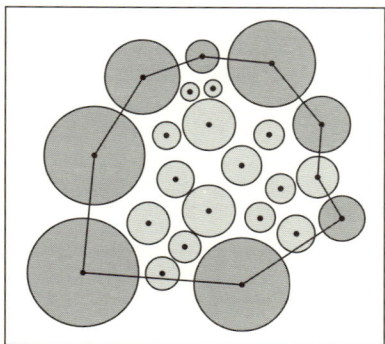

Fig. 10.4. The Coins Problem: a solution. Illustration by Ali Nesin.

(a) The noose is just the boundary of the convex hull of the centers of coins.

(b) When using the hammer, try to minimize the noise.

(c) Remove the nail before you attempt to move the coin.

I will return later to the discussion of the crucial methodological question: why should we debase mathematics and formulate the problem in terms of coins and its solution in terms of cords and nails, instead of formulating it in proper mathematical language:

> *Given a finite collection of non-intersecting convex compact sets in $\mathbb{R}^2$ ...* [?]

Yes, indeed, reword the problem in a formal language.

Meanwhile I continue the story about the further adventures of the problem.

The success of his problem moved Eugene to rework the problem for a higher-level competition, the All-Union Mathematical Olympiad.

The Convex Coins Problem. There are a finite number of coins on the table, each shaped as a convex polygon (as in Britain). The coins do not touch each other. Prove that it is possible to slide the coins along the surface of the table to its edge, one by one, without clicking a coin against another. [?]

Solve it without reading further.

And here events took quite an unexpected turn. When Khukhro offered the problem to the meeting of the organization committee of the All-Union Olympiad, Sergei Konyagin (who was a trainer of the national team) looked into his briefcase and took out a sheet of paper with the following formulation.

The City of N-sk Problem. In the city of N-sk, all buildings are direct prisms over convex N-gons (of course, we assume that there are only finitely many buildings). Prove that, no matter from what direction a traveller approaches the city, at least one of the buildings is not eclipsed (even partially) by others. [?]

Solve it without reading further.

Unfortunately, I cannot give a reference, but I vaguely remember that the problem was published in *Kvant* magazine 25 or 30 years ago.

The reader understands, of course, that the problem immediately lost its sporting value (serious mathematics competitions use only new problems).

But the story continued. At a mathematical conference my colleagues celebrated the publication of the book *Mathematical Aquarium* [292] written by one of the conference speakers, Victor Ufnarovski. I leafed through a copy of the book and, in astonishment, discovered one more version of the Post Office Conjecture.

The Soap Bubbles Problem (Victor Ufnarovski). Someone makes soap bubbles (each of a spherical shape). The bubbles float

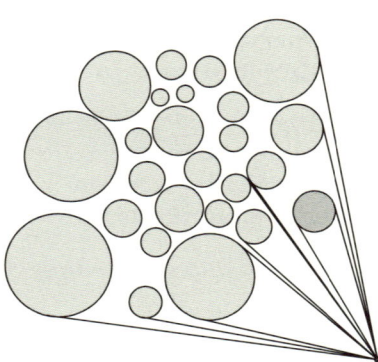

Fig. 10.5. A solution of the Soap Bubbles Problem. Illustration by Ali Nesin.

*Solve it with-
out reading
further.*
around the room. Prove that at least one bubble is not eclipsed by the others in the vision of a given observer. [?]

Ufnarovski was extremely surprised to hear my story about the parcels. He came to his problem actually making soap bubbles, with his child, and had never seen similar problems before.

There is one more similar problem, from the fascinating book *Lines and Curves* by Gutenmacher and Vasilyev [268]. I was reminded about its existence only in 1989, eleven years after I formulated my Post Office Conjecture.

The Holes in the Cheese Problem. A big cubic piece of cheese has some spherical holes inside (like Swiss Emmental cheese, say).
*Solve it with-
out reading
further.*
Prove that you can cut it into convex polytopes in such way that every polytope contains exactly one hole. [?]

10.2 Solutions

I will give some very brief solutions, starting with the Soap Bubbles Problem. The answer can be given in just two lines:

> The sphere with the shortest tangent line connecting it to the point of view is not eclipsed (Figure 10.5).

The two-dimensional version of the Soap Bubbles Problem gives, of course, a new solution to the Round Coins Problems (Figure 10.6). The City of N-sk Problem is also not that difficult after you have analyzed the solution of the Soap Bubbles Problem. Indeed, it is quite clear that the crucial role should be played by the *supporting* (or "tangent") lines from the point T which marks the position of

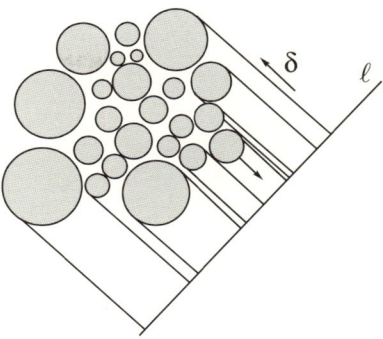

Fig. 10.6. Another solution of the Round Coins Problems. Illustration by Ali Nesin.

the traveller, to the buildings. Recall that everything takes place in the plane; therefore you can draw two supporting lines from T to any building: the left line and the right line, which touch the left and the right corner of the building, as seen by the traveller. Therefore the solution can again be formulated in one sentence:

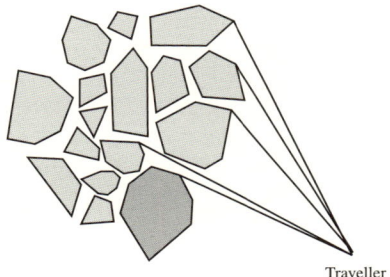

Traveller

Fig. 10.7. The City of N-sk Problem: a solution. Illustration by Ali Nesin.

Of all right corners visible choose the leftmost; it belongs to a building which is not eclipsed by the others (Figure 10.7).

Finally, observe that the statement of the Holes in the Cheese Problem, if true, gives us a solution of the Soap Bubbles Problem. Indeed, imagine that the soap bubbles are holes in the huge transparent piece of cheese. Let us cut the cheese into convex polytopes, each containing exactly one hole. Then the bubble which happens to be in the same polytope as the observer is, of course, not eclipsed by any other bubble.

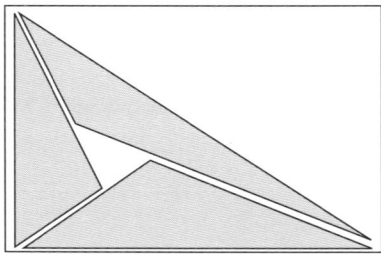

Fig. 10.8. For the Holes in the Cheese Problem. Illustration by Ali Nesin.

Another useful observation is that if the holes are not spherical, but just convex, then the conclusion of the Cheese Problem is no longer true: it is fairly easy to produce a counterexample even in the planar case (Figure 10.8). Therefore we come to a very important conclusion that the metric properties of the sphere—they distinguish the sphere from arbitrary convex bodies—are relevant. This leads to the idea of reusing the trick with the shortest tangent line and yields the following working conjecture:

> The set of points P such that the tangent line from P to the given sphere is not longer than the tangent lines from P to the other spheres is a convex polytope. [?]

Prove it!

And what happens when we have only two holes? Then polytopes become half-spaces (because their union is the entire space) and our working conjecture transforms itself into a rather plausible statement:

> The set of points P such that the tangent lines from P to two (non-intersecting) spheres are equal is a hyperplane (which, obviously, separates the spheres). [?]

Prove it!

The last statement can be easily proven; all you need is the Pythagorean Theorem and some simple properties of a tangent line to a circle. The resulting plane is known in geometry as a *radical plane* of two spheres.

The proof, of course, can be easily reduced to the planar case. The notion of the *radical axis* makes sense for intersecting circles as well. Moreover, it is easier in that case, since the radical axis is just the straight line through the two intersection points. The software package CINDERELLA [303] provides a very nice illustration of this fact: you start with intersecting circles and the line defined by their points of intersection and gradually decrease the radii of the circle. Eventually, the circles disengage but the line does not disappear! And it is exactly the radical axis, this time of non-intersecting circles. Actually, the Euclidean plane as it is implemented in CINDERELLA is the real affine part of the complex projective plane; the radical axis of two non-intersecting circles is the real part of the line through two *complex* points of intersection; since the intersection points of two real circles are complex conjugates, the line is invariant under complex conjugation and therefore shows up on the real Euclidean plane. This brings to mind Hadamard's quip: *The shortest route between two truths in the real domain passes through the complex domain.*

After our short excursion into circle geometry the Cheese Problem can be handled without any further trouble:

> For every pair of holes in the cheese we construct its radical plane and take the intersection of all half-spaces determined by the radical planes and containing the given hole. The resulting convex polytopes form the desired partition of the piece of cheese.

When the holes are just points (that is, our spheres have radius 0), the resulting partition of the space into convex polytopes is a well-known mathematical construction. It is called the *Voronoi diagram* of the set of points, and the polytopes are known as *Voronoi cells*. In the Cheese Problem we had a generalization of a Voronoi diagram: here, *Voronoi cells* are defined by comparing distances to a closest hole. Voronoi diagrams have a surprising range of applications, from optical character recognition to materials science, where they are used to approximate polycrystalline microstructures in solid media [392].

Reformulation, translation of problems from one mathematical language to another, is virtually unknown not only in high school mathematics, but at the college and undergraduate level as well.

10.3 Some philosophy

We handled the problems with flair and ease, but this feeling is misleading. Taken individually, the problems are much more difficult. We make our task easier by using the key ideas of the solution of one problem as heuristic tools for attacking other problems.

Some of my readers, I hope, are professional mathematicians and know how useful it can be to work not with one problem in isolation, but with a variety of related versions, or reformulations. Alas, reformulation, translation of problems from one mathematical language to another, is virtually unknown not only in high school mathematics, but at the undergraduate level as well.

The search for an adequate language for the distillation of the essence of a problem is one of the key components of mathematical work.

Also, it greatly helped us that we understood the shared mathematical content of all the problems. The mathematician reader will easily find a uniform formulation in terms of convex geometry. But do the problems become easier in the abstract formulation?

My story has approached the point which shows its purpose. The Post Office Problem is a very good model of one of the key components of mathematical work—the search for an adequate language for the distillation of the essence of a problem. In fact, it is irrelevant whether the words of this language are mathematical terms or ordinary words of common, everyday language. What matters is whether they capture the intuitively felt relations between objects involved in the problem.

In mathematics teaching, we need to demonstrate the "linguistic" aspect of mathematics to our students—and as early as possible.

I recalled the валенки story when, almost 15 years after formulating the Post Office Conjecture, I got involved in a project in combinatorics and started to think about a generalization of the concept of convexity suitable for use in purely algebraic situations. I suddenly realized that my struggle was not much different from a schoolboy's attempts to find words for the description of the notion "the coin is outside of the rest of the coins". The notions "outside", "out-most", and "outsider" were so simple but became very difficult as soon as the schoolboy tried to consider various cases of mutual positioning of coins. It was a typical Zasetsky's torture; see Section 7.1.

The analogy extends much further—in our toy problems we have two examples of the use of already well-known languages, namely, of order and ordered sets ("the rightmost of all left corners") and metric geometry ("the shortest tangent"). Of course, the language of ordered sets appeared in disguise, but only because our everyday language is very efficient in manipulating concepts such as "left–right". But it does not change the nature of our discoveries; in "real", "research" mathematics we use the machinery of already existing theories in essentially the same way. Again, in "big" mathematics everything starts with the search for an adequate language.

Also, let us recall why we decided to use the metric properties of the sphere. We generalized the problem from spherical holes to convex holes, compared with another generalization (round coins–convex coins) and realized that, since the generalization does not work, we have to use specific (that is, metric) properties of the sphere. By the way, this is a typical example of a *meta-argument*, when we look at a mathematical theory from outside and use mathematics or mathematical logic to understand its essential features.[2]

I think that, in mathematics teaching, we need to demonstrate this "linguistic" aspect of mathematics to our students—and as early as possible. One should not think, however, that the reformulation of mathematical theories in "everyday" language can replace systematic study. Our little problems about coins are not convex analysis!

The problems considered in this chapter can be traced to four absolutely independent sources. But their authors belong more or less to the same mathematical school and share some common understanding of the basic structures of mathematics.

It is a very interesting question to what extent the authors were guided by the well-known and documented structures and concepts of mathematics (in this particular case, by the formal definition of a convex set), and to what degree by informal knowledge which is not documented in books or journal papers and is passed mostly from teacher to student by direct contact.

It might be a sweeping generalization on my part, but I think that a mathematician, in his work, uses, first of all, informal interpretations of his theory and uses formal language to record his informal findings. The story of the Post Office Problem shows that each mathematical school has its shared systems of informal interpretations of the key concepts used by the school. The main secret of learning mathematics at the "research" (say, MSc or PhD) level is to master these informal aspects. How this is done remains a mystery to me.

> *A mathematician first uses informal interpretations of his theory, and only then switches to formal language to record his informal findings.*

10.4 But is the Post Office Conjecture true?

So, it is time to return to the Post Office Conjecture. Alas, it is not true. The first counterexample was offered by Kuzminykh a few years after the conjecture was formulated. In this chapter, I have been trying to avoid the use of pictures (to emphasize the point that usual everyday language turns into mathematics when it allows us to express, clearly and precisely, everything which we want to say, and without recourse to gestures and pictures). But this is the point where we possibly need a picture.

Take a convex polytope with obtuse angles between adjacent faces (for example, an icosahedron). Place, next to each face, a thin

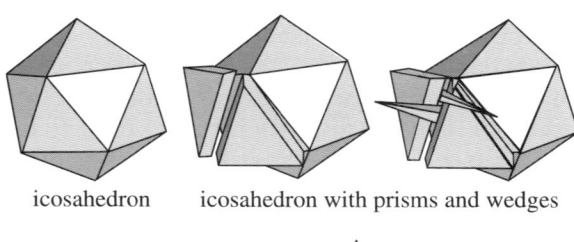

icosahedron   icosahedron with prisms and wedges

a wedge

Fig. 10.9. The Post Office Conjecture: a counterexample. Illustration by Ali Nesin.

right prism with a base of the same shape as the face, and insert, in a criss-cross pattern, thin wedges, each wedge being a right prism over a triangle with an obtuse angle (Figure 10.9). It is now obvious that we cannot remove any polytope from this heap.

This solution is nice, of course, but, unfortunately, it appeals to the same kind of geometric intuition which tells us that Rubik's Cube is impossible. The problem is much subtler than one may think after being prompted by Kuzminykh's counterexample. If your intuition tells you that the counterexample is, indeed, correct and refutes the Post Office Conjecture, I will now kill your confidence.

Theorem (Guennady Noskov, circa 1988). A heap of convex bodies can be disassembled if we are allowed to move all bodies *simultaneously*. Can you see why this is true in Kuzminykh's example? [?]

> *Prove it without reading further!*

Noskov's proof is strikingly elegant.

Apply to the heap a homothetic transformation with coefficient $1+\lambda$, starting with $\lambda = 0$ (the identity transformation), with λ growing to infinity. Then the bodies will expand, and the heap will also expand, retaining its configuration and the relative position of all bodies. Inside every body we can move, by means of a parallel translation, the original copy of the body. [?]

> *Noskov's Theorem can be generalized from convex to star-shaped bodies—check!*

In more popular terms, the solution can be described as follows. Imagine that each body is covered by a thin elastic membrane, a kind of balloon. Let us pump air into all the balloons. The balloons will start to expand, pushing each other, and inside each balloon we can move its body. Simple?

Alas, we reached the point where, it appears, the expressive power of everyday language is exhausted. Why do the balloons push each other apart? Will the result be different if we inflate different balloons at different rates? When you start thinking about that question, the "balloon" solution ceases to be self-evident. The homothetic transformation used by Noskov in his solution ensures an equal rate of inflation of each balloon.

Well, we should not be disappointed that everyday language does not work any longer at the apex of our little theory. It is natural; like poetry, the very reason for the existence of mathematics is that it expresses thoughts and feelings which we cannot express in mundane everyday language.

And the final comment: as Alexander Kuzminykh has recently explained to me, he also was not the first one to find a counterexample to the Post Office Conjecture. He discovered in [339, pp. 141–143] a reference to a paper by Fejes Toth and Heppes [350] which exhibited an arrangement of 14 convex bodies in $\mathbb{R}^3$ none of which can be moved independently of the rest. Moreover, Danzer and also Dawson gave examples with twelve congruent convex bodies. See Dawson's paper [262].[3] Apparently, Noskov's Theorem is also not new; it appeared in [262, Theorem 3].[4] There is nothing surprising in this: the circle of problems on convexity related to the Post Office Conjecture is natural and beautiful, and it is only natural that many different people have set them up and found solutions. Finally, the Post Office Conjecture is about disassembling heaps of convex bodies using *one hand*.

Şükrü Yalçınkaya, aged 4.5

If the use of *two hands* is permitted, we get a wonderful (and practically important, in view of applications to robotics) problem, solved by J. Snoeyink and J. Stolfi in [405].

10.5 Keystones, arches, and cupolas

> *Did ye never read in the scriptures,*
> *The stone which the builders rejected*
> *the same is become the head of the corner.*
> Matthew 21:42

The whole story is not that new: the ability of convex bodies to interlock has been exploited, for many centuries, in architecture and building practices in the design of arches and cupolas. The stone at the top of the arch, the one that locks the whole construction, has been given special, frequently mystical, treatment—hence the proverbial *the head of the corner* or *keystone* of the Bible.

Fig. 10.10. *From* Wikipedia*: Columns of Ramses III at Medinat Habu. Public Domain.*

As with all human inventions, this was not always the case: the technology was unknown in Ancient Egypt, and Egyptian temples were forests of columns (Figure 10.10). Even at later times, architects appear to have been hesitant to use keystones and *voussoirs*, other wedge-shaped stones of the arch, and preferred more clumsy *false arches* (Figure 10.11).

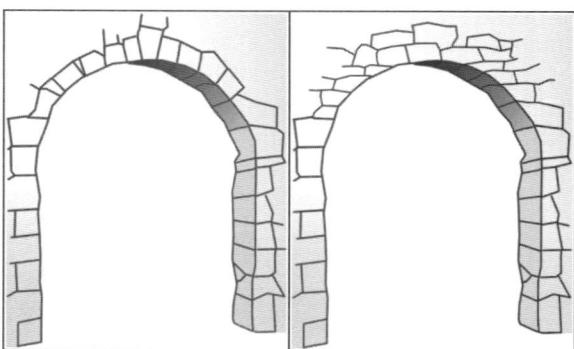

Fig. 10.11. *Comparison of (left) a generic "true" stone arch and (right) a corbel (or false) arch. Source:* `http://en.wikipedia.org/wiki/ Image:Arc\_truefalserp.jpg,` *distributed under Creative Commons Attribution ShareAlike 2.5 License.*

Even more fascinating examples of arches can be seen in two of my pictures from Turkey. In the first one, Figure 10.12, voussoirs at the entrance to a *mescit* (small mosque, chapel) of Agžikarahan, a Seljuk caravansaray on the Silk Road, Cappadocia, are traditional convex wedges.

Fig. 10.12. Mescit (a small mosque) in Agžikarahan, a Seljuk caravansaray on the Silk Road, Cappadocia. Photograph: A. V. Borovik

In another photograph, Figure 10.13, of the Main Gate of Agžikarahan, voussoirs are elaborately jaded, as if the builder did not trust the plain wedge shape to support the weight of the magnificent construction.

Alexey Kanel-Belov proposed an interesting version of the Post Office Problem:

Design an arch bent *downwards* and composed of convex stones. [?]   Indeed, try it!

[A solution is suggested in the next section.]

Fig. 10.13. *Ağžıkarahan, Main Gate. Photograph: A. V. Borovik*

10.6 Military applications

> Броня крепка и танки наши быстры . . .

Remarkably, one can place congruent Platonic solids in an interlocked regular pattern [346, 347, 348]; see Figure 10.14. [?] Since some of the hardest materials in nature come in the form of, say, octahedral crystals and, moreover, can be manufactured only as relatively small crystals, the possibility of an interlocked regular pattern leads to the idea of making composite materials from separate, but interlocked, crystals. There is another benefit to this approach: failure of materials is in many cases associated with the propagation of macroscopic fractures. But a material made of very hard interlocked crystals is pre-fractured and therefore developing faults will be trapped in the existing gaps between crystals and their further propagation will be arrested by the mutual interlocking of crystals. Hence the Post Office Conjecture could potentially lead to the development of a promising type of composite materials.

Spatial configurations of interlocked octahedra are paradoxical and beautiful. But it is Noskov's expansion procedure (see page 208) which is the true marvel. You can reverse it and make it into a manufacturing process for the mass production of new composite materials: place crystals in appropriately positioned nests in

An interlocked pattern in Figure 10.14 is made of cubes; sketch a similar pattern made (a) of octahedra and (b) of tetrahedra.

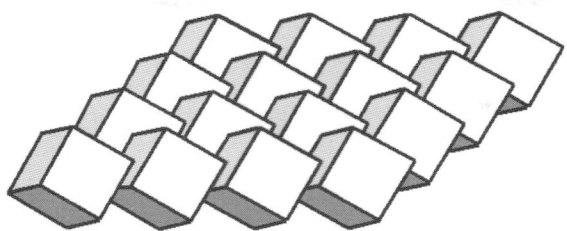

Fig. 10.14. Interlocked cubes in a regular pattern after Dyskin et al. [348].

a low density foamy material (something like styrofoam), at comfortable distances one from another, like glass Christmas tree decorations in their protective packaging. Then pour in some solvent and compress the foam, until the crystals reach their interlocked positions.

Meanwhile, Alexey Kanel-Belov tried to promote the idea for use in tank armor (I do not know whether he succeeded or not). It was a rather unexpected turn of events for my problem about валенки. Kanel-Belov asked the crazy question: can the interlocked heap be made from regular solids—and, indeed, he found appropriate patterns. However, he was not the first one to come up with the idea—a couple of years before him, some Dutch road builders described a design of a very sturdy foundation layer for road surfacing made from interlocked concrete tetrahedra [355].

Fig. 10.15. *Autoportrait* by Ali Nesin.

Notes

[1]Actually, валенки are not boots but very thick felt socks; they have no seams. Also, валенки are not socks because the felt is so thick that they are rigid and stand up like, say, Wellingtons. Unlike Wellies, which are to some degree foldable, you cannot fold a валенок.

To a Briton, felt footwear without a rubberized sole might seem to be an inane idea; however, you have to take into consideration that, in a Siberian winter, both air and snow are exceptionally dry.

[2]MATROIDS. I mentioned my work in combinatorics. To be precise, I recalled the валенки story when I was trying to reformulate an important combinatorial concept—matroid—in terms of convexity (in some very general understanding of this word); here, I avoid precise definitions. I wish only to note that the convexity of matroids happened to be related to the following property: from wherever you look at the matroid, it always has a right and a left corner (after you have defined what is meant by "look", "left-right" and "corner"). The idea of this approach to matroids is fully developed in my book with Israel Gelfand and Neil White [320].

[3]INTERLOCKED CUBES IN A TORIC ARAANGEMENTS. Igor Pak wrote to me:

> After some thinking about Dawson's and Kanel's results I realized that one can make an amazing new example of a solution to the Post Office Conjecture with unit cubes! Simply, make cubes a little bit smaller to give them some room of freedom. Everything is robust and still locked. Then use the regular pattern and start bending it a little bit eventually to glue it together into a huge torus (embedded into $\mathbb{R}^3$) with a finite number of non-removable polytopes. Similarly, one can take an example from dodecahedra (use a base construction as in Dyskin-Kanel-...). I believe that in fact any convex polytope can be used, although never bothered to prove that. This should be contrasted with Dawson's theorem that spheres cannot be locked this way.

[4]And possibly in Satz 1 (Satz von De Bruijn) of Fejes Toth and Heppes [350].

Part III

History and Philosophy

11

The Ultimate Replicating Machines

Be fruitful, and multiply ...
Genesis 1:28

Returning to the Davis-Hersh definition of mathematics as "the study of mental objects with reproducible properties" (Section 2.4), so far I have been concerned mostly with various interpretations of the word "mental". Let us now look at "reproducible", which appears to be another key word.

In this chapter, I intend to venture into *memetics*.

Memetics is an emerging interdisciplinary area of research concerned with the mechanisms of the evolution of human culture. The term *meme* was made popular[1] by Richard Dawkins [167] and was introduced into the mainstream philosophy and cultural studies by Daniel Dennett [25]. It refers to elementary units of cultural transmission and invokes the concept of *gene*. The word "meme" is a recombination of the Greek word *mimēma* which means "that which is imitated" and "gene". Memes play the same role in the explanation of the evolution of culture (and the reproduction of individual objects of culture) as genes do in the evolution of life (correspondingly, the reproduction of individual organisms). Although Susan Blackmore's book *The Meme Machine* [155] has been translated into twelve (maybe even more by now) languages and made memetics a recognizable discipline, memetics still fights for its place among other sciences and is slowly morphing from a great metaphor into a theory. Indeed, this is what scientific theories do in the course of their development: they start as brilliantly colored butterflies of metaphors and paradoxes and then turn themselves into dull caterpillars and spend the rest of their lives munching, day and night, their statistical tables. So far, I have not seen serious statistical tables in papers on memetics.

Indeed, the concepts of "meme" and "meme complex" still look more like beautiful metaphors rather than rigorously defined sci-

217

entific terms. They are too general and require further specification when applied to particular cultural systems. There is nothing wrong with this state of affairs; the same observation can be made about the theory and practice of genetic or evolutionary algorithms in computer science [366, 384, 387]. Here, specific case studies and applications (like the one described in [378]) are much more interesting than a rather vacuous general theory.[2]

I am probably not the first person to apply Dawkins's concept of memes to mathematics.[3] Indeed, this appears to be a natural and obvious line of discourse. But obvious things are not always interesting; in my discussion of memes as units of the transmission of mathematics, I will try to concentrate on non-trivial aspects of the meme metaphor.

However, my GOOGLE searches and perusal of the *Journal of Memetics* suggest that no one has so far looked seriously at the memetics of mathematics. Filling this gap is the primary aim of this chapter.

As I attempt to argue, mathematical memes play a crucial role in many meme complexes of human culture: they increase the precision of reproduction of the complex thus giving an evolutionary advantage to the complex, and, of course, to the memes themselves. Remarkably, the memes may remain invisible, unnoticed for centuries and not recognized as rightly belonging to mathematics. I argue that this is the characteristic property of "mathematical" memes:

> If a meme has the intrinsic property that it increases the precision of reproduction and error correction of the meme complexes it belongs to and if it does that without resorting to external social or cultural restraints, then it is likely to be an object or construction of mathematics.

This chapter is addressed to two disjoint groups of readers: mathematicians and memeticists. This makes it difficult to set the right level of mathematical detail. Memetics is in a state of flux, and it is very difficult to assess the composition of the memetics research community. Still, I have a feeling that the area is dominated by researchers with a background in the humanities. I will be happy if my book is read by some fellow mathematicians. But I also thought that it would be prudent to adjust the technical level of discussion and explain some facts of mathematical practice in more detail.

11.1 Mathematics: reproduction, transmission, error correction

*If you concentrate too closely
on too limited an application of a mathematical idea,
you rob the mathematician of his most important tools:
analogy, generality, and simplicity.
Mathematics is the ultimate in technology transfer.*
Ian Stewart [290]

At this point I hope that the reader comes prepared to agree that the rigor of mathematics depends on the reproducibility of its concepts and results. The very fact that we can talk about mathematical constructs as *objects* stems from this reproducibility. Let us look at a simple example of the "objectification" of mathematical constructs as discussed by Davis and Hersh [21, p. 407] (see also the discussion of "reification" in Section 6.1):

I do not know whether 375,803,627 is a prime number or not, but I do know that it is not up to me to choose which it is.

Here, a *prime number* is a natural number that is not the product of two numbers different from 1 and itself. For example, 6 is not a prime number because $6 = 2 \times 3$, while 5 is a prime number. The question about whether 375,803,627 is a prime number is harder, but it still has a definite answer of "yes" or "no". I can choose which numbers I "like" and which I don't, which numbers are "lucky" and which are not, but it is not up to me to choose whether 375,803,627 is prime or not, because if I carry out the necessary arithmetic computations or run standard software to find it out, my conclusion would be exactly the same as that of any other person who made similar computations and avoided arithmetic errors. Moreover, if someone applies to the number 375,803,627 some clever mathematical theorem—and does not make errors in his or her arguments— the conclusion will again be exactly the same. Notice that all these caveats: "correctly", "avoided errors", etc., refer to the same reproducibility, to the fundamental fact that across the mathematical community, there is a common understanding of what kind of argument or computation is "correct" and what is not.

Therefore mathematics studies mental objects with reproducible properties which happen to be built according to reproducible rules, with the precision of reproduction being checked by specific mechanisms, which, in their turn, can also be reproduced and shared. I hate to complicate the picture further but must mention that these rules can themselves be treated as mathematical objects (this is done in a branch of mathematics called mathematical logic) and are governed by metarules, etc. We come to an even more interesting observation: mathematical objects can reproduce themselves only because they are *built hierarchically*. Simple or atomic objects (definitions, formulae, elementary arguments, etc.) form more complicated entities (theorems and their proofs) which, in their turn, are arranged into theories.

11.2 The Babel of mathematics

> *And the whole earth was of*
> *one language, and of one speech.*
> Genesis 11:1

In the previous section, we came to the conclusion that mathematical truths do not exist on their own but only in their interaction with an extremely elaborate and sophisticated web of mathematical concepts, constructions, and results. This web of mathematics is probably one of the most complex constructions ever built by humans.

People outside the mathematical community cannot imagine how big mathematics is. Davis and Hersh point out that between 100,000 and 200,000 new theorems are published *every year* in mathematical journals around the world. A poem can exist on its own; although it requires readers who know its language and can understand its allusions, it does not necessarily refer to or quote other poems. A mathematical theorem, as a rule, *explicitly* refers to other theorems and definitions and, from the instant of its conception in a mathematician's mind, is integrated into the huge system of mathematical knowledge. In my analysis, I stop here and do not consider how mathematics is submerged into the wider social and cultural environments. The interested reader can find a very interesting discussion of the social nature of mathematics in Davis and Hersh [21] with further development in the book by Hersh [48].

We notice next that mathematics is a product of an *evolutionary development* in the work of many generations of mathematicians. Moreover, it continues to evolve (for a non-mathematician, a good snapshot of an evolving mathematical theory is the famous book by Lakatos [64]). The wide spread of the phylogenic tree of mathematics is reflected in the extreme variety of mathematical disciplines and areas of specialization. The universally accepted Mathematics Subject Classification scheme contains about 5,000 entries. For a mathematician, it is perfectly possible to build a successful career working strictly within the boundaries of a single one of these 5,000 mathematical disciplines—from his or her PhD thesis via papers and books to the supervision of the PhD theses of his/her students. If you take two papers at random, say, one on mathematical logic

Fig. 11.1. *The Tower of Babel*, by Pieter Bruegel the Elder, 1563. Kunsthistorisches Museum, Vienna.

This image is in the public domain because its copyright has expired. This applies to the United States, Canada, the European Union and those countries with a copyright term of life of the author plus 70 years.

Why did mathematics not share the fate of the builders of the Tower of Babel?

And the Lord came down to see the city and the tower, which the children of men builded.

And the Lord said, Behold, the people is one, and they have all one language; and this they begin to do: and now nothing will be restrained from them, which they have imagined to do.

Go to, let us go down, and there confound their language, that they may not understand one another's speech. (Genesis 11:4–7)

and one on probability theory, you may easily conclude that they have nothing in common. However, a closer look at the Mathematics Subject Classification scheme reveals discipline 03F45: "probability logics and related algebras".

We see that, despite all this diversity, there is an almost incomprehensible unity of mathematics. It can be compared only with the diversity and the unity of life. Indeed, all life forms on Earth, in all their mind-boggling variety, are based on the same mechanisms of replication of DNA and RNA.

Can this analogy be explained by mathematics being based on highly specific and precise mechanisms of reproduction of mathematical memes?

Like genes in a biological population, memes replicate within an evolving cultural system and "propagate themselves in the meme pool" [167, p. 192]. In that respect, what makes mathematics special is that it possesses powerful mechanisms which ensure the precision of replication. When comparing mathematics with other cultural systems, we see that a high precision of replication can usually be found in systems which are relatively simple (like fashion, say). In other cases the precision of reproduction is linked to a certain rigidity of the system and an institutionalized resistance to change, as in the case of religion. We do not offer hecatombs to Zeus, but, after 2,000 and something years, we still use Euclidean geometry—and this has happened without anything resembling the endless religious wars of human history.

Mathematics is so stable as a cultural complex because it has extremely powerful intrinsic capability for error detection and error correction. As Jody Azzouni puts it [6]:

> What makes mathematics difficult is (1) that it's *so easy* to blunder in; and (2) that it's *so easy* for others (or oneself) to see—when they're pointed out—that blunders *have* been made.

I recommend Azzouni's paper to the reader; it analyzes, in considerable detail, the difficulty of explaining the astonishing power of self-correction of mathematics by external factors, social or cultural. I claim that the only possible explanation lies in the nature of the mathematical memes themselves.

Susan Blackmore [155] stresses the role of imitation in the propagation of memes. Most mathematicians would be uncomfortable about the emphasis on imitation as a mechanism for propagation of mathematical memes. As we shall see, the mechanisms of reproduction of mathematical memes are much more precise and specific.

11.3 The nature and role of mathematical memes

I was looking for a simple case study—and could not believe my luck when I found the following example in Richard Dawkins's *Foreword* [168] to the book by Susan Blackmore [155].

> Memes travel longitudinally down generations, but they travel horizontally too, like viruses in an epidemic. [...] Crazes among schoolchildren provide particularly tidy examples. When I was about nine, my father taught me to fold a square of paper to make an origami Chinese junk. [...] The point of the story is that I went back to school and infected my friends with the skill, and it then spread around

the school with the speed of the measles and pretty much the same epidemiological time-course. [...] My father himself originally picked up the Chinese Junk meme during an almost identical epidemic at the same school 25 years earlier. The earlier virus was launched by the school matron. Long after the old matron's departure, I had reintroduced her meme to a new cohort of small boys.[4]

At this point, I feel that it could be useful to include diagrammatic instructions for making a Chinese Junk (Figures 11.2, 11.3).

Meanwhile we return to Dawkins. He goes on analyzing the case of the Chinese Junk meme.

A favourite objection to the meme/gene analogy is that memes, if they exist at all, are transmitted with too low fidelity to perform a gene-like role in any realistically Darwinian selection process.

He draws a distinction between low-fidelity and high-fidelity memes by describing a thought experiment where a child is given a picture of a junk and asked to copy it in his own drawing, and then the next child is asked to copy the drawing of the first child, and so on. Of course, we expect that the pictures will very soon deteriorate beyond recognition. However, the skill of making a junk is passed from a child to child with high fidelity, if, in a different experiment, we ask a child to teach another child how to fold the junk. As Dawkins put it,

...inheritance in the drawing experiment is Lamarckian (Blackmore calls it 'copying-the-product'). In the origami experiment it is Weismannian (Blackmore's 'copying-the-instructions'). In the drawing experiment, the phenotype in every experiment is also the genotype—it is what is passed on to the next generation. In the origami experiment, what is passed to the next generation is not the paper phenotype but a set of instructions for making it. Imperfections in the execution of the instructions result in imperfect junks (phenotype) but they are not passed on to future generations: they are non-memetic.

So, we are already in the close vicinity of the Davis-Hersh "mental objects with reproducible properties". But have we arrived at mathematics? Look for yourself at Dawkins's description of

...the Weismannian meme line of instructions for making a Chinese junk:
1. Take a square sheet of paper and fold all four corners exactly into the middle. (This makes a *blintz*, in the terminology of English origamists.)

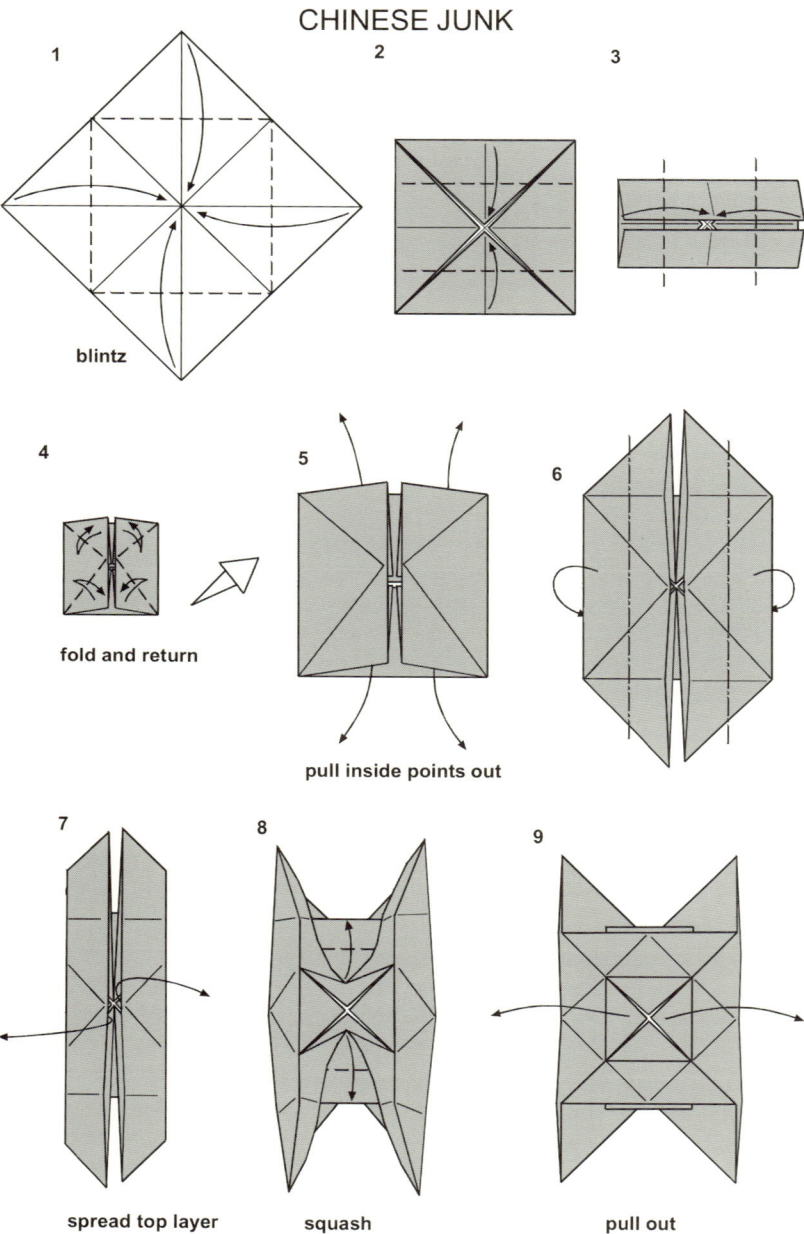

Fig. 11.2. Diagrammatic instructions for making the Chinese Junk, part I. Diagram © D. Petty, reproduced with his kind permission.

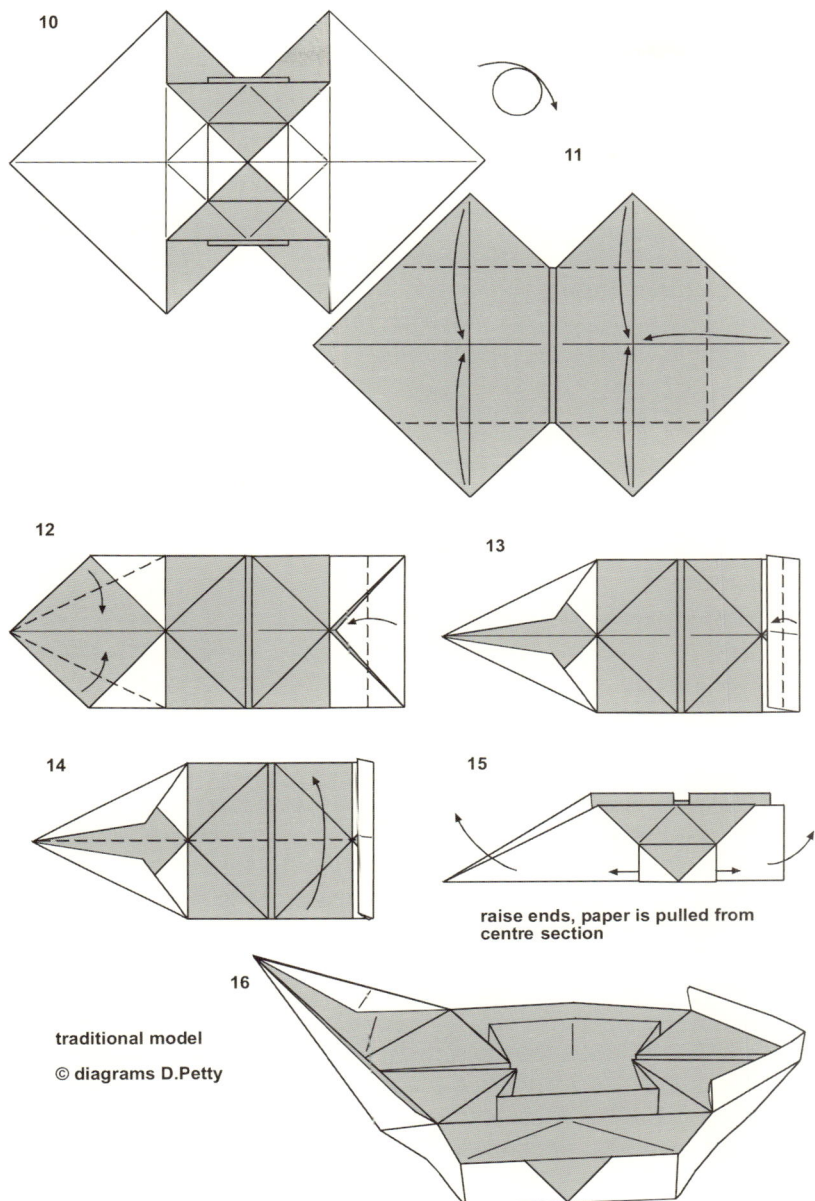

Fig. 11.3. Diagrammatic instructions for making the Chinese Junk, part II. Diagram © D. Petty, reproduced with his kind permission. For more on Origami, see [432, 433].

> 2. Take the reduced square so formed, and fold one side
>   into the middle ...

[The interested reader may find the rest of the instructions, in a diagrammatic form, in Figures 11.2 and 11.3—AB.]

Yes, it is mathematics. It is exactly the mathematical nature of the procedure which ensures its high resistance to errors. Moreover, it generates a remarkable ability to self-repair errors within the elementary steps of the procedure. Let us look at step 1. In effect, it requires finding the center of the square. This can be done in surprisingly many ways, always leading to exactly the same result. For example, this is one way to fold:

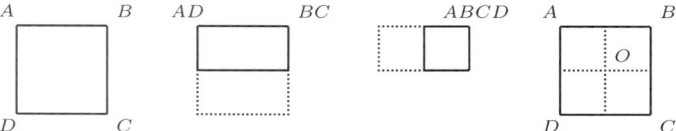

and this is another:

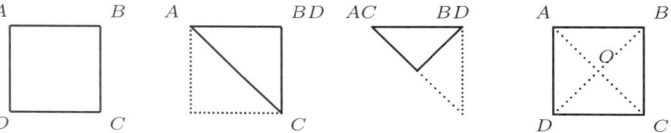

After the center is found, there is only one way to fold the given corner into the center; it is a mathematical fact (and a theorem of geometry: in the Euclidean plane, for any two distinct points A and O, there exists exactly one line l such that the axial, or mirror, symmetry in l brings A to O). Therefore there is simply no way to introduce an instructional error at that step.

Please notice that, in this simple example, we have a primitive version of *cryptomorphism*: the center of the square, treated as a mathematical object on its own, can be defined (or constructed) in many different ways, but it remains exactly the same object, and, in subsequent use in further constructions, it becomes irrelevant how we got it in the first instance. As we shall soon see, the concept of cryptomorphism is of crucial importance for understanding the evolution of mathematics.

This example neatly explains the nature and the role of mathematical objects: in the evolution of human culture, they are memes which happened to be successful and which spread because of the following properties:

- They have extreme resilience and precision of reproduction.
- When included in meme complexes (collections of memes which have better chances for reproduction when present in the genotype as a group), they increase the precision of reproduction

of the complex as a whole. This property is so important that it deserves a special name; I suggest the term *intrinsic error-correction facility* and will call memes with this property *correctors*.

- I stress that the error-correcting property of mathematical memes is *intrinsic*; they do so without resorting to external social or cultural restraints. (It is tempting to add here: "without resorting to the reproducible and stable features of the natural (physical) world". This is a delicate point of bifurcation of mathematics and physics; I discuss it in more detail in Section 11.6.) Therefore I exclude from the consideration such remarkably reproducible entities as religion and law.

- The property of cryptomorphism is a bonus: as we have seen in a simple example, it helps to correct possible errors in reproduction.

Remarkably, memes may remain invisible, unnoticed for centuries and not recognized as rightly belonging to mathematics. But I argue that intrinsic error correction is the characteristic property of "mathematical" memes, and I wish to formulate the following thesis as a challenge to memeticists:

THE CORRECTOR MEME THESIS.
Whenever you encounter an efficient corrector meme, a closer look at it should reveal a hidden object, procedure, or construction of mathematics—even if the mathematical nature of the meme has gone unnoticed for centuries.

Refute my thesis—I would be most happy to have a go at the analysis of your counterexample. But before you try to beat me in that game, have a look at the analysis of another example, of truly historic dimension, the square grid method invented in Ancient Egypt for copying drawings to murals (Section 11.5).

To conclude our discussion of corrector memes, I wish to suggest that it is natural to expect that if corrector memes form meme complexes, the latter should be very stable and reproducible. And these, in effect, were coacervate drops in the primeval soup of human culture which started the evolution of mathematics.

However, if you feel really uncomfortable with my Corrector Meme Thesis, you may wish to skip the following discussion and go directly to Section 11.6 where I discuss various caveats and possible complications.

11.4 Mathematics and Origami

I have to warn the reader that, unlike the previous material in this chapter, this section will build up a steep mathematical learning curve.

The mathematics of paper folding as discussed in Section 11.3 is not that naive: actually, the simple foldings which allow one to find the center of the square already contain in themselves the seeds of a complete axiomatic development of Euclidean geometry. One should not underestimate the astonishing power even of the simplest mathematical memes.

As Figure 11.4 shows, folding is the same as the "flipping" of images which our brain does, according to Pinker and Tarr (Section 2.2), in the process of recognition of flat mirror images. It is something which is very easy to visualize; it is one of the most self-evident concepts of geometry. But, as we shall soon see, foldings give much more than a convenient intuitive interpretation of symmetry. [?]

> *When we fold a sheet of paper, why is the fold line straight?*

As a geometric concept, folding is exactly what its name suggests: the plane $\mathbb{R}^2$ is being folded on itself like a sheet of paper. The unit vector α normal to the fold line l gives the direction of folding. Notice that this introduces the clear distinction of the roles of the two half-planes H^+ and H^-.

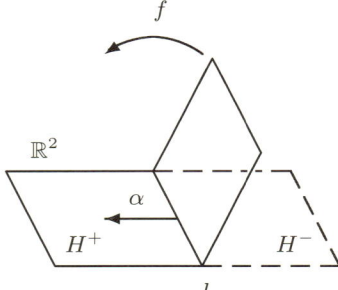

Fig. 11.4. Folding.

One way to demonstrate the expressive power of paper foldings in the formal axiomatic development of geometry is to observe that every construction of Euclidean geometry done with a compass and a straightedge can be done just by folding paper.

The proof of this fact is given by Roger Alperin [253, 254].[5] His treatment of the problem is algebraic: he uses the interpretation of points in the plane with Cartesian coordinates (x, y) as complex numbers $x + yi$. As Figure 11.5 illustrates, all the usual geometric constructions can be interpreted as the construction of points (complex numbers), starting with the fixed distinguished points 0 and 1.

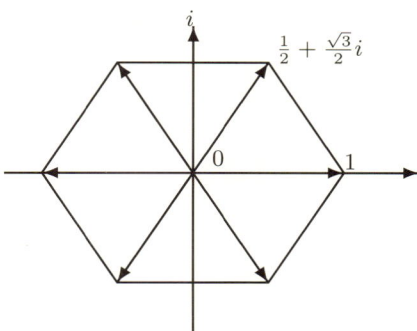

Fig. 11.5. To construct the regular hexagon with side 1 means to construct, starting with two distinguished given points 0 and 1 on the complex plane $\mathbb{C}$, the point $\sqrt[6]{1} = \frac{1}{2} + \frac{\sqrt{3}}{2} i$.

Alperin gives the following iterative procedure which leads to the class of *constructible* complex numbers. The basic objects are *points* and *fold lines* or *creases*.

(1) The line connecting two constructible points is a fold line (therefore, if we start with two points 0 and 1, the very first fold is the line through 0 and 1, the real line of the complex plane).
(2) The point of intersection of two fold lines is constructible.
(3) The perpendicular bisector of the segment connecting two constructible points can be folded.
(4) The line bisecting any given constructible angle can be folded.
(5) Given a fold line l and constructed points A, B, then whenever possible, the line through B that reflects A onto l can be folded.

I omit one more axiom since the first five already allow us to carry out every compass-and-ruler construction [253, Theorem 4.1].

[?]

Fold a regular hexagon.

These five axioms allow for the construction of any complex number which belongs to the *Euclidean field*, that is, the number field which can be characterized as the smallest field containing the rational numbers and which is closed under taking all square roots. All numbers which can be constructed by compass and ruler belong to the Euclidean field (this result, in its modern form, is due to David Hilbert, although its history can be traced back to François Viète [95, p. 371]). Hence every ruler-and-compass construction can also be done by paper folding!

Observe that axiom (5) above contains the words "whenever possible" which does not allow us to treat Alperin's axioms as self-contained; an exact description of the scope of the applicability of axiom (5) is easy to find: the set of all points which can be obtained from the given point A by reflecting in fold lines passing through B is, obviously, the circle centered at B of radius $|AB|$ (indeed, if A' is the reflection of A in a line through B, then, obviously, points A and A' are equidistant from B: $|AB| = |A'B|$). This description, however simple, still refers to properties of Euclidean geometry not contained in Alperin's axioms.

It is not very difficult to modify axiom (5) and to make it self-contained. Why not try the following axiom?

(5°)   Given a fold line l and constructible points A, B outside of l, if there is a fold in a line through B which brings A onto the other side of l from B, then the line through B that reflects A onto l can be folded.

Roger Alperin,
aged 13

Notice, however, that axiom (5°) refers to the fact that a fold line divides the plane in two halves. This property is most natural in the context of Origami geometry: indeed, there are two foldings in the given line, determined by which half of the sheet of paper remains on the table and which one is turned over.

This obvious distinction between the points which are moved by folding and which "stay on the desk" allows us to define the relation "to be between": we say that a point C *lies between* the points A and B if C stays on the desk whenever A and B stay on the desk. Can you see how Alperin's axioms imply that the point C in that case actually lies on the fold line connecting A and B? In conventional terms of Euclidean geometry this means that C belongs to the segment with the endpoints A and B.

The only catch here is that Euclid famously missed, in his axiomatization of geometry, the betweenness relation, which was added to the axioms of Euclidean geometry in the 19th century.[6] As we see, betweenness and order appear in Origami geometry in a most natural way; it is much harder to miss it.

Alperin's axioms were designed as a *description* of the construction process, not for *proving* theorems within a formal axiomatic framework. For example, to check that various ways of finding the center of the square discussed in Section 11.3 all yield *the same point*, we have no choice but to invoke facts from Euclidean geometry, which therefore retains its status as the ambient structure for all Origami constructions.

In order to make these Origami axioms of Euclidean geometry absolutely self-contained, one needs to add a few more axioms to Alperin's axioms. This will bring them closer to Bachmann's axiomatization of Euclidean geometry in terms of symmetry [309] and to Gustave Choquet's axiomatic approach to Euclidean geometry which prominently uses axial symmetries [295]. Choquet even uses, in one of the versions of his axioms, foldings as a primary concept [295, Axiom IV', Appendix 1]. For more details, see Section 11.8.

11.5 Copying by squares

The mathematics in this section is once again elementary. Rather grand mathematical terminology is mentioned—but it is not actually used.

So, I repeat my claim that the role of mathematical memes in the meme complexes of human culture is to ensure high fidelity reproduction.

Things become a bit confusing when we start to think about reproduction being done by a machine. For example, every time you listen to a CD player, a sophisticated mathematical algorithm checks for possible errors and corrects every packet of digits the player reads from the CD;

The role of mathematical memes in the meme complexes of human culture is to ensure high fidelity reproduction.

other algorithms then convert the digits into music. The error-correction part is crucial for ensuring the stable quality of reproduction of a record. So-called *error-correction codes* are developed and analyzed in the branch of mathematics called coding theory. Error-correcting algorithms are hardwired into millions (possibly billions, if you take account of mobile phones) of microchips and are now one of the most ubiquitous uses of mathematics. But these codes are too deeply immersed in the machine technology and so too invisible in everyday human-to-human interactions to fall within the scope of this chapter.

However, there are examples of codes specifically designed for correcting human errors. Perhaps the most ubiquitous is the ISBN, International Standard Book Number, found on the cover or at least on the copyright page of almost every book. The ten symbols of the ISBN bear in themselves a small mathematical device which allows one to detect (single occurrences of) the two most common errors people make in typing: substitution of one digit for another and transposition of two digits. (The old version of the ISBN used the alphabet consisting of the digits $0, 1, \ldots, 9$ and the extra symbol X; each ISBN contained 10 symbols. The new version has 13 digits.)

See the discussion of the ISBN in Ray Hill [301] or in David Poole [304] and the discussion of the use of a similar code in a Norwegian census [288]. Alternatively, if you would like to be a crypt-

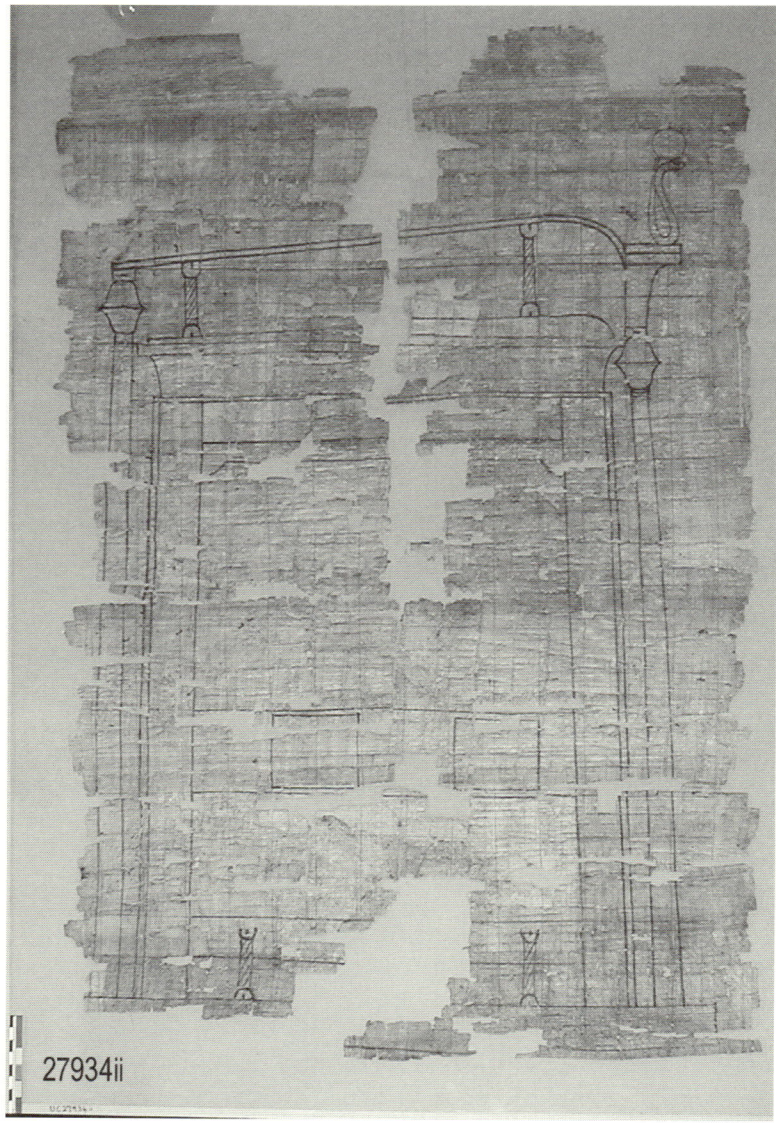

27934ii

Fig. 11.6. Fragments of an unfinished mural with the square grid of red guidelines, Ancient Egypt, Old Kingdom. © Petrie Museum of Egyptian Archaeology, University College London; item UC27934, reproduced with permission. The grid spanned not just the drawing but the entire world:

> The mathematical framework that underlines the structure of his artwork
> is but one aspect of the Egyptian concept of space. The Egyptian was highly
> conscious of the box-like structure of his world, traversed by two co-ordinates
> at right angles: the generally south-north flow of the Nile, and the east-west
> passage of the sun across the ceiling of the heavens, which was supported by
> the third axis. The contiguous planes of this environment are carefully de-
> fined as separate entities and are to be found in the fully developed Egyptian
> temple, which is strictly cubic and is a model of the universe at its creation.
> [422, p. 13]

analyst, collect ISBN codes from every book you have and try to figure out how they work. [?]

In [304], pp. 52 and 55, you will also find the description of two other ubiquitous error-correcting codes: the Universal Product Code associated with bar codes on merchandize and Codabar system for credit card numbers.

So, if the example of the Chinese Junk Origami has not convinced you that "corrector" memes are mathematical objects and procedures in disguise, then take a look at a photograph of an unfinished Ancient Egyptian tomb mural (Figure 11.6), with the square grid of guidelines still visible. The transfer, square by square, of the image from a sketch to the wall, is a mathematical procedure. This is possibly the earliest example of the use of mathematics in copying. I cannot imagine a better illustration of the principle that the role of mathematical memes in the meme complexes of human culture is that mathematics (even if it remains invisible) ensures high fidelity reproduction. Even more remarkable, it is more *copying-the-product* in the sense of Blackmore than *copying-the-instructions*.

The notion of copying square by square reappeared thousands of years later, in mathematical crystallography and in the theory of Lie groups, and nowadays goes by the grand name of the *action of a discrete subgroup of a Lie group on the set of translates of its Dirichlet region* (or *fundamental domain*, which is the same). In the copying-by-squares example, the discrete subgroup is the set of all translations of the plane which preserve the grid; the Lie group is the set of all translations of the Euclidean plane); the Dirichlet regions are the squares of the grid.

One of the more elementary examples of this set-up can be found in Minkowski's treatment of the Dirichlet Units Theorem from algebraic number theory. It may be useful to put Minkowski's work into a wider mathematical context and then explain how it is related to copying by squares. The so-called *Minkowski Lemma* is a classical mathematics competition problem but also (in a slightly more general *n*-dimensional version) appears, for example, as Lemma 1 in Section II.4.2 of the respected treatise on number theory [319]. The fact that it bears its creator's name is a reflection of its importance. The Minkowski Lemma is used in the theory of integer solutions of certain classes of underdetermined polynomial equations with integer coefficients. One of the simplest examples is, say,

$$x^2 - 3y^2 = 6.$$

This particular equation has infinitely many solutions, and the Dirichlet-Minkowski method allows one to find all of them. (Try

> *A symbol X which occasionally appears in ISBN codes, but only in the rightmost position, is a hint.*

Ray Hill,
aged 6

to find a couple of solutions. Can you describe a rule which lists them *all* in a systematic way, without trying every pair (x, y)?) The Minkowski Lemma, which I will soon state, was used by Minkowski to justify that a certain method of solving equations indeed gives *all* solutions.

It is also worth mentioning that the historic roots of the *Diophantine equations*—the umbrella name for the class of polynomial equations with integer coefficients—can be traced back to Ancient Egypt and Babylon; another and possibly independent origin of the concept was in Ancient India. It was then and there that the remarkable discovery about the triangle with sides 3, 4, and 5 (the famous Pythagorean triangle, $3^2 + 4^2 = 5^2$) was made: namely, that it had a right angle. There are suggestions that a rope with $3 + 4 + 5 = 12$ knots tied on it at equal measures (assuming that the rope is a closed loop!) could be conveniently used for dividing fields into rectangular plots.[7]

And here is the statement of the Minkowski Lemma:

> Assume that a slip of colored paper of unknown shape but with area smaller than 1 is placed on a sheet of graph paper with squares of area 1. Then it is possible to slide the colored slip without rotating it and in such way that in its new position it does not cover any of the nodes of the graph paper. [?]

Prove the Minkowski Lemma without reading further!

As you will see in a second, the proof of the Minkowski Lemma uses the Pigeonhole Principle (also known as the Dirichlet Principle; see Section 4.1), with the important modification that it is applied not to numbers of elements of finite sets, but to areas of geometric figures.

We start the proof of the Minkowski Lemma by observing that, instead of moving the colored slip, we may as well try to find a way to move the grid of the graph paper (without rotating it) in such a way that, in its final position, no node of the grid is covered by the slip.

How will we find this movement of the grid? Let us pause for a second and not touch the grid. Instead, cut the slip along the grid lines and move each bit into the same square S, preserving the relative position of the colored bit in its square. Now all bits and pieces of the original colored slip are in the same square; since their total area is less than 1, they cannot cover the whole square. Now we can describe the desired movement of the grid: slide one of the nodes of the grid into a "free" point of the square S, keeping the grid lines parallel to themselves.

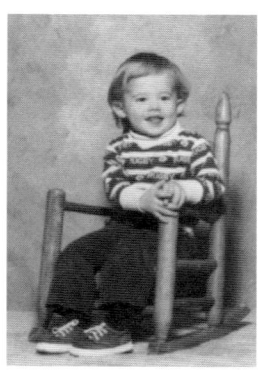
Jeff Burdges, aged 3

As we see, this remarkable proof uses "copying by squares" in its original form, going back in time all the way to Ancient Egypt.

It would be timely to return to the Corrector Meme Thesis: are any doubts left that copying by squares is mathematics?

11.6 Some stumbling blocks

To be on the side of caution, I have to warn the reader about possible stumbling blocks. The first two are fairly obvious and do not threaten my thesis in any serious way.

11.6.1 Natural language and music

I exclude natural human languages (and alphabets!) from the scope of my thesis—they have to be treated as the ambient medium of human culture.

Music appears to be extremely resilient in reproduction, too; but a closer look at it shows that it is saturated by mathematical structures (we cannot even say "hidden" since the link was very obvious to Pythagoras, more than two millennia ago). See the excellent book *Music and Mathematics* [32] for more detail on the intertwining of music and mathematics, or David Benson's treatise [9], or a recent book by David Wright [143].

However, other aspects of the concept of mathematical memes require some attention.

11.6.2 Mathematics and the natural sciences

One may reasonably ask the question of whether there are memes which maintain fidelity merely through reproducibility of experiments and not through mathematics. Some elementary steps in culinary recipes, for example, like "bring to boil", are there simply because boiling is a very convenient and easy to reproduce reference point. Cooking slowly at $85°$C quite frequently produces much juicier and tastier results, but the right conditions are much harder to control. To demarcate mathematics from physics[8] means to treat the reproducibility of human culture differently from the way we use and study the stability and reproducibility of the laws of the physical world. I accept that the line between physics and mathematics is hard to draw; however I have no choice but to refrain from further discussion if we are not to get stuck in the classical philosophical conundrums of relations between (ideal) mathematical objects and the real (physical) world.[9] Even in Dawkins's Chinese Junk study, these philosophical questions are already present: the folding of a piece of paper could be viewed as an entirely physical process, and the reproducibility of the properties of the center of the square as an entirely physical phenomenon.[10] On the other hand, the words "square" and "center of the square" apparently refer to ideal concepts, creations of the human mind. I hope that the reader is prepared to accept that such stable and reproducible entities as "the boiling point" and "the center of the square" have quite different mechanisms of reproducibility.

Our brief discussion, however, was not a waste of time; indeed, it has prompted us to formulate an important warning: the memetics approach to mathematics does not eliminate or resolve the classical philosophical problems of mathematics. We may only hope that memetics can perhaps usefully reformulate some of the old conundrums.

11.6.3 Genotype and phenotype

Another difficulty which we have to address was already present in the Chinese Junk but became much more prominent in another case study, that of copying by squares (Section 11.5). When a much bigger meme complex benefits from the error-correcting properties of a particular mathematical meme, these properties could be viewed as expressed in the phenotype. The ISBN error-correcting code has the self-correcting "memotype" of being mathematically provable, which is very similar to the self-correcting "memotype" of the Chinese Junk, but the ISBN also has the "phenotype" of correcting book orders. It is worth noting that mathematicians are concerned more with the intrinsic properties of mathematical memes than with their expression in phenotype. I may anticipate some very interesting technical discussions of the subtle distinction between "internal" and "external" error-correcting facilities, but I am not prepared to go into any detail in this book.

A much more interesting question is the difference between memes already recognized as part of mathematics and memes which have not yet been recognized as such. Are high fidelity memes reborn as objects of mathematics only when they are discovered by mathematicians and described in an acceptable mathematical language? It is difficult to avoid the impression that, in the discussion of the genotype/phenotype distinction, we revisit the classical dispute between the Platonists and formalists on the nature of mathematics: are mathematical objects discovered or created?

11.6.4 Algorithms of the brain

The concept of mathematical memes bears the seeds of yet another complication: it is quite possible that the reproducibility of certain basic mathematical concepts and algorithms is determined more by biology than culture.

The reader certainly noticed that the "flipping" and "swooping" described by Pinker (Section 2.2) are essentially the same as "folding", which was used, in the previous sections, as a basic mathematical concept to generate Euclidean geometry. Therefore "folding" provides an example of a mathematical concept which is directly rooted in the subconscious neurophysiological mechanisms of the human mind. This makes the analysis of the nature of "atomic" objects and concepts of mathematics even more exiting.

11.6.5 Evolution of mathematics

The evolution of mathematics is an interesting topic, since, by their very nature, mathematical memes have virtually zero mutation rate. We can talk only about the evolution of meme complexes and about the change in relative frequency of particular memes in comparison with

The mathematical core of the theory does not evolve: it emerges gradually from the fog.

their cryptomorphic analogues (that is, differently expressed memes with essentially the same mathematical content).

For example, there is no intrinsic mathematical reason why Euclidean geometry was not developed in the language of Origami; the traditional language of Euclidean geometry and the language of Origami are *cryptomorphic*; that is, they are mathematically equivalent and only superficially different. The Origami language for geometry stood no chance because the ancient Greeks had no paper.

Evolution and the resulting shape of the language of a mathematical theory is influenced by external social and cultural factors. At the risk of sounding like a Platonist, I tend to think that a mathematical theory should be understood as the totality of all its potential cryptomorphic reformulations, including those in current use, as well as those used in the past and now forgotten, or even those which have not yet been invented. Therefore the mathematical core of the theory does not evolve: it emerges gradually from the fog.

Further discussion of these exciting topics, however, goes well beyond the scope of the present book.

One obvious conclusion, however, can be recorded. When you compare Origami geometry (pretty obscure) with traditional Euclidean geometry (quite glorious), you realize that memetics should treat value judgements (so prominent and prevalent in mathematics) as one of the many factors of social selection; but memes, as such, bear no intrinsic value.

Richard Booth,
aged15

11.7 Mathematics as a proselytizing cult

Many features of mathematics as a cultural (and even social) system can be explained by the nature of mathematical memes. I would compare mathematics with beekeeping: the stereotypical beekeeper is a very mild-mannered gentleman, and this is predetermined by biological, not social, factors: bees would not tolerate, for example, a jumpy hyperactive boy messing around their beehive—as I know from my own childhood encounters with my uncle's bees.

In a similar way, mathematics shapes mathematicians—we are meme-keepers of mathematical memes, and they behave rather differently from other memes of human culture. Since my book is about mathematical practice and not mathematicians, I will mention only two specific traits, relevant to the subject of cultural transmission.

First of all, one should not underestimate the proselytizing zeal of mathematicians (which should not be found surprising since mathematicians are selected by their urge to disseminate mathematical memes). But this is a separate story which I may write about elsewhere.

Another trait which I cannot leave without a brief discussion is the mathematicians' obsession with standards of typography. Many mathematicians are paranoid about keeping tight control over their texts; one has to accept that they are driven by the urge to minimize the errors of transmission.

Mathematics set the aesthetic standards of printed text at a level unheard of before. Is it really surprising that TeX, the best and most powerful and flexible typesetting system [58], was designed by a mathematician? The mathematician reader will immediately recognize that the present book is set with LaTeX, a version of TeX and the de-facto standard of typesetting in mathematics. A French artist Bernar Venet has even used formulae from mathematical books typeset with Donald Knuth's TeX system as objects of art, projecting them onto walls and incorporating them into murals (Figure 11.7). We give two quotations from Hofmann [53] who drew the attention of the mathematical community to Venet's art:

> Donald Knuth first empowered us—and by now forced us—to typeset our own texts and create the typography of our formulae with our own hands. While it is true that we are doing this with our fingers on the keyboards rather than by assembling lead cast letters, Knuth has shaped his programming language TeX so that it faithfully emulates the original craft.
>
> Having created a fine typographical product thus adds extra satisfaction to the pleasure of having found a mathe-

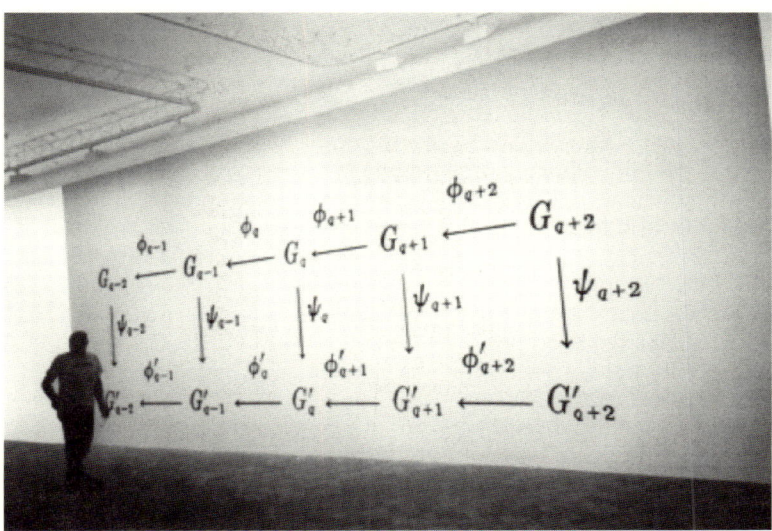

Fig. 11.7. Bernar Venet, *"A homomorphism of an exact lower sequence"*. 2000, wall painting. Centre d'Art Contemporain Georges Pompidou. Cajac, France. Reproduced with permission, © Bernar Venet.

matically aesthetic result, having proved it, and having presented it in a stylistically elegant fashion.

A reviewer of this book reminded me that one has to remember the period before Knuth. The ugliest books since the invention of printing were those typewritten mathematics books of the 1970s with handwritten symbols. Mathematicians gladly embraced the photocopying of badly typed manuscripts because the previous technology, manual typesetting of mathematical formulae, was expensive, slow, and exceptionally prone to error. It is enough to quote one of the many diatribes from Littlewood's *A Mathematician's Miscellany* [65, p. 38]:

> A minute I wrote (about 1917) for the Ballistic Office ended with the sentence: 'Thus σ should be made as small as possible'. This did not appear in the printed minute. But P. J. Grigg said, 'what is that?' A speck in a blank space at the end proved to be the tiniest σ I have ever seen (the printers must have scoured London for it).

Should we be surprised then that, despite the tiny print runs, mathematical publishing remains a profitable line of business because the publishers benefit from mathematicians' masochistic self-exploitation—driven by our memes, we are prepared to take upon ourselves the toil and cost of typesetting our books and papers.

11.8 Fancy being Euclid?

This section is more technical and can be skipped.

What amendments are needed for turning Alperin's axioms for Origami geometry (Section 11.4) into a self-contained Origami axiomatic system of Euclidean geometry?

- We need a formal description of the properties of folds which would allow us to distinguish between two "opposite" folds.
- We must postulate some combination of two opposite folds as, in effect, a reflection.
- Reflections preserve all the properties of our geometry; they are *automorphisms*.
- Fold lines and their points of intersection satisfy the axioms of an affine plane.
- We need some kind of "betweenness" axiom.
- Finally, we have to formalize the fact that we have plenty of folds (Alperin's axioms (3)–(5) do exactly that).

The resulting axiomatic system will have considerable redundancy; I believe that by excluding or modifying the axioms one by one and analyzing the resulting theories, it is possible to make a compact and beautiful system of axioms.

I offer this problem to the reader as an exercise. Actually, it is a good team project in a course on foundations of geometry—if such courses, once a compulsory part of the training of mathematics teachers (at least in Germany and Russia) are still taught anywhere in the world.

Here is my attempt at a possible set of axioms for Euclidean geometry in terms of Origami.

An *Origami geometry* is a set E of elements (called *points*) and a family $\mathcal{F}$ of maps from E to E (called *folds*) which satisfy the following axioms:

Origamy I. *Axioms for folds*:

   I(i) Folds are idempotent maps, $f \cdot f = f$.

   I(ii) For every fold f there exists a unique *opposite* fold $-f$ such that the set-theoretic union of functions

$$\sigma_f = f \mid_{-f(E)} \cup -f \mid_{f(E)}$$

   is an involutory (that is, $\sigma_f^2 = \mathrm{Id}$, the identity transformation) one-to-one map from E onto E; we call it the *reflection* or the *flip* associated with the fold f.

   (Flips turn the sheet of paper over, in such a way that the corresponding fold line remains at precisely the same position on the desk.)

   I(iii) If f is a fold, then the set of fixed points of σ_f is called the *fold line* of f and is denoted $\mathrm{line}(f)$. We demand that the unordered pair of folds $\{f, -f\}$ be uniquely determined by the fold line $\mathrm{line}(f)$.

I(iv) Flips send fold lines to fold lines and folds to folds: if σ_f is a flip and g is a fold, then $\sigma_f \cdot g \cdot \sigma_f$ is a fold, whose line is $\sigma_f(\text{line}(g))$.

The last axiom says that flips are *automorphisms* of our geometry. In more elementary language, this means that a fold sends another fold line to a fold line or a union of two half-lines.

As you can see, this first group of axioms is exceptionally dull—as frequently happens when you formalize self-evident properties of intuitively clear mathematical objects.

Origami II. The second group if axioms is essentially Alperin's, slightly strengthened by inserting the *uniqueness* requirements for foldings.

II(i) For any two distinct points there is a unique fold line containing these points.

II(ii) Two fold lines intersect in at most one point.

II(iii) For any two distinct points A and B there is a unique fold which sends A to B.

II(iv) If l and n are two fold lines, then there is a flip which sends l to m.

II(v) For this axiom we need to introduce the concept "being on the same side of the fold line l". Let f and $-f$ be two folds associated with the fold line l. We say that points A and B lie on the *same side* of l if either $f(A) = A$ and $f(B) = B$ or $-f(A) = A$ and $-f(B) = B$; otherwise we say that A and B lie on the *opposite sides* of l.

With this definition at hand, we demand that, given a fold line l and points A, B outside l, if there is a fold in a fold line through B which brings A onto the other side of l from B, then there is a fold line through B, such that one of its two folds maps A onto l.

Origami III. *Parallel Lines*. We say that fold lines l and m are *parallel* if they either coincide or have no points in common. We demand that "being parallel" is an equivalence relation.

Without this axiom, we have no way to distinguish between Euclidean and hyperbolic geometries, the latter also having an Origami version. Still, it may make sense to uniformly develop the common part of Euclidean and hyperbolic geometries by omitting the Parallel Lines Axiom.

Origami IV. *Betweenness*. We say that a point C *lies between* the points A and B if every fold which does not move points A and B also does not move C. The *segment* $[A, B]$ is the set of points which lie between A and B.

We demand that if points A and B lie on the opposite sides of a fold line l, then the segment $[A, B]$ through A and B intersects l.

Notice that Axioms II(i), II(ii), and III mean that points and fold lines form an affine plane. To tie everything together in one rigid structure, I invoke the least intuitive and the most powerful of all the axioms on my list, *Bachmann's Axiom*. It is an affine version of the key axiom in Bachmann's axiomatization of Euclidean geometry in terms of symmetries [309]. It would be nice to replace Bachmann's Axiom with something more elementary and self-evident. On the other hand, folds are mappings; why should we shy away from the abstract framework of sets and mappings?

Origami V. *Bachmann's Axiom*. Let $\sigma_l, \sigma_m, \sigma_n$ be the flips corresponding to fold lines l, m, n. The identity

$$(\sigma_l \sigma_m \sigma_n)^2 = \mathrm{Id}$$

holds if and only if all three lines l, m, n have a point in common or are parallel to each other.

Origami VI. *Non-degeneracy*. A geometry has at least two distinct points.

How do you get points not collinear with the first two?

To start an interesting geometry, we usually need four points in "general position", that is, any three of them not belonging to the same line. But, given just two distinct points, we can use folds to breed more. [?]

It is very likely that this system of axioms is highly redundant and can be replaced by a much more compact group of axioms. For example, I believe that Bachmann's Axiom can be replaced by the following *Unstretchability Axiom*. In its formulation, a *motion* is a composition of several flips; a *half-line* is the set of points on a line fixed by a fold which moves some points on the line non-trivially.

Origami V°. *Paper is unstretchable*. Consider an *angle*, a figure consisting of a point A and two distinct half-lines l and k emanating from A. A motion which fixes A and maps l into l and k into k is the identity map.

I have not checked all the details, but I have reason to believe that these axioms (perhaps with some technical adjustments) suffice to prove that points and lines form an affine plane over some ordered commutative field K where you can extract square roots of positive elements and that the flips are orthogonal symmetries with respect to some anisotropic symmetric bilinear form on the underlying vector space. As I have already mentioned, I leave this problem as an exercise for the reader.

Of course, it is possible to treat the theory in an entirely synthetic way (that is, without introducing numbers and coordinates). The interested reader may wish to develop axiomatic Origami geometry up to the following theorem: the three altitudes of the triangle intersect in a common point.[11].

For obvious reasons, perpendiculars (hence altitudes in triangles) are likely to play a larger role in Origami geometry than in the traditional treatment of Euclidean geometry. Vladimir Arnold once mentioned, in passing [104], that the three altitudes theorem follows from the Jacobi identity in Lie algebra. It is likely that Arnold meant the cross product algebra of vectors in three-dimensional Euclidean space, or, in algebraic terms, the Lie algebra of the orthogonal group $SO_3(\mathbb{R})$.[12] In vector (cross product) notation the Jacobi identity looks like this:

$$(\vec{a} \times \vec{b}) \times \vec{c} + (\vec{b} \times \vec{c}) \times \vec{a} + (\vec{c} \times \vec{a}) \times \vec{b} = \vec{0}.$$

Indeed, Arnold's remark was converted into an elegant proof by Skopenkov [404]. Therefore the three altitudes theorem is truly fundamental; it makes a sensible focus point for the theory.

What is the basis of my confidence that a purely axiomatic development of Origami geometry is feasible? A more abstract group-theoretic approach to Euclidean geometry, centered on symmetry rather than on folds (which is, essentially, the same) is well known: the relevant theory was developed by Bachmann [309]. I state the key result of Bachmann's book in the form due to Schröder [401].

> Assume that a group G is generated by involutions (that is, elements of order 2). Assume further that the set I of all involutions of a group G possesses the structure of a projective plane in such a way that three involutions i, j and k are collinear if and only if their product ijk is an involution. Then G is isomorphic to the special orthogonal group $SO_3(K, f)$ for some field K of characteristic $\neq 2$ and a nonisotropic quadratic form f on K^3.

I once unknowingly rediscovered Bachmann's axiomatization of Euclidean geometry, working from scratch: I needed an abstract description of the group of rotations of three-dimensional Euclidean space in terms of half-turns (rotations through $180°$ about an axis). The Euclidean rotation group keeps appearing as a phantom "minimal" configuration in my work in model-theoretic algebra. Some details can be found in [321, Chapter 8].

Of course, we know that a half-turn in three dimensions is just an axial symmetry in a plane containing the axis. Moreover, this axial symmetry has, in this context, a most natural mechanical interpretation, as turning the plane over, as a sheet of paper—which is the combination of two opposite foldings with the same fold line.

Therefore the axiomatics of Euclidean geometry via Origami should be feasible, and this is why it can be safely assigned to the playground of fledgling young mathematicians. But it really helps to know a bit of group theory and to work at a higher level of abstraction—even if the final streamlined version of the axiomatic system and the corresponding theory is formulated in very, very elementary terms: just paper and foldings.

Notes

[1]THE ORIGINS OF THE TERM "MEMETICS". I wrote in earlier versions of the text that the word "memetics" was coined by Richard Dawkins, but

Jonathan Vos Post corrected me: *"Meme was not coined by Dawkins. When I took Psycholinguistics at Caltech in 1972 or 1973, I was already able to quote the word MEMEME from existing specialized literature.*

[2]My own interest in memetics and in the evolution of mathematical objects is motivated by my work on genetic algorithms [317, 318]. The work done by my co-authors and myself was a case study, and a very peculiar one: we traced the co-evolution of a population of (non-deterministic) evolutionary algorithms for a particular mathematical problem to a *deterministic* mathematical algorithm.

[3]GOOGLE searches for "meme" and "mathematics" lead to very disappointing results, mostly texts which use the word "meme" in a derogatory fashion, with the meaning "a silly belief spread by contagion". Keith Devlin in his MAA online column, `http://www.maa.org/features/ invisible.html`, gives, as an example of a "fully developed and well established mathematical meme," the jingle *all mathematicians have no sense of humor.*

[4]David Lister, an authority on the history of Origami, claims that Chinese Junk was known in Europe since the early 19th century: in 1806 a clear drawing of this paper folding was made in Holland. See Figures 11.2 and 11.3 for diagrammatic instructions for making a Chinese Junk.

[5]AXIOMS FOR ORIGAMI. Discussion of axioms for Origami can be found in Hartshorne [299, Exercises 28.12–28.15] and Holme [302, Section 16.8]. Thomas Hull kindly informed me that the development of axioms for Origami geometry has a long history which can be traced back to a paper of 1936 by Margherita Piazzolla Beloch [258]. In 1992, a comprehensive list of axioms for Origami was given by Humiaki Huzita [281].

[6]A detailed discussion of "betweennes" and its history, with extensive bibliographic references, can be found in Coxeter [296, Section 12.2].

[7]ROPES WITH KNOTS are mentioned in Stillwell [91, p. 2]; also, see discussion of the cuneiform text *Plimpton 322* in [96] and [79, 80]. Some authors, however, fiercely dispute the actual use of the rope with knots.

[8]MATHEMATICS AND PHYSICS. Many prominent mathematicians will view yet another attempt to separate mathematics from physics with utter contempt.

[9]MATHEMATICS AND THE "REAL WORLD". The philosophical conundrums of relations between (ideal) mathematical objects and the real (physical) world become even more mind-boggling when you observe that mathematical logic studies something real: reason. This comment is due to Boris Zilber.

[10]ORIGAMI AND TOPOLOGY. The (mathematical!) study of the topological properties of Origami by Eric Demaine et al. [340, 341] sheds some light on relations between the physical environment and mathematical abstraction of Origami. It addresses in Origami problems the subtle

> *distinction between specifying the geometry of the final folded state (a single folding, e.g. an Origami crane) and specifying a continuous folding motion from the unfolded sheet to the final folded state (an entire animation of foldings)*

and proves that every "good" folded state can actually be reached.

[11]THE THREE ALTITUDES THEOREM is not part of Euclid's *Elements*. A drawing of three altitudes is found in one of the works by Archimedes, *Book*

of Lemmas [1]; see Hartshorne [299, p. 52] for a discussion. It is likely that the three altitudes theorem can be proven by an appropriate adaptation of Hjelmslev's "calculus of reflections" (see [299, Theorem 43.15]; originating in Hjelmslev's paper of 1907 [363], the calculus of reflections was developed into an impressive theory by Bachmann [309]).

[12]CROSS PRODUCT. The cross product algebra of vectors in three dimensions is the Lie algebra of the Lie group $SO_3(\mathbb{R})$, a fact of fundamental importance, say, for mechanics. It still surprises me that this is not found in standard undergraduate textbooks.

12

The Vivisection of the Cheshire Cat

Why has not man a microscopic eye?
For this plain reason, man is not a fly
Say what the use, were finer optics giv'n,
T' inspect a mite, not comprehend the heav'n?
Alexander Pope, *Essay on Man*

12.1 A few words on philosophy

I wish to conclude my book by explaining my position with respect to the principal issues of the philosophy of mathematics. In brief, my standpoint can be formulated in three bulleted points:

- First and most of all, I do what most practicing mathematicians do—I try my best to avoid awkward questions. This is why I make a distinction between *mathematics* and *mathematical practice;* I do that in the hope that this simple verbal trick allows me to escape uncomfortable discussions on the philosophical problems of mathematics at least temporarily.
- Secondly, whenever possible, I try to supplant philosophical questions with metamathematical problems. Of course, this is what philosophers have done for the past century. My position differs on two counts: I look at much more "local" problems, and, in search of answers, I go beyond set theory and mathematical logic into more "concrete" areas of mathematics.
- Finally, I try to further narrow the field of the philosophical problems of mathematics by moving some of them into the rapidly expanding realm of cognitive science.

In this chapter, you will find examples of all three types of arguments. The second and third are of special interest to me; I would like to see how much philosophy is left in mathematics after a systematic separation from metamathematics and cognitive science.

Still, I have to address some basic questions arising from my understanding of mathematics as the "study of mental objects with reproducible properties". (I am grateful to David Corfield who brought potential complications to my attention.) The reader has probably noticed that I also freely talk about the "reproduction of mental objects".

It is natural to ask what is being reproduced, an object or a property? Is a mathematical object nothing more than its properties?

I have to admit that I, like many (if not most) mathematicians, do not care about the precision of metalanguage used in discussing mathematics—because our primary language is already sufficiently precise. Mathematicians working in the same field frequently talk to each other about their extremely technical work using very loose language—yet they understand each other perfectly. This is possible only because their loose talk refers to a shared formal framework. On the other hand, few things irritate a mathematician more than a seminar talk on an unfamiliar subject if the speaker avoids giving explicit definitions or statements of results, indulging instead in an "ideological" discussion.

The previous chapters should already have made it clear that, for me, an object is an encapsulated sum of its properties and functions; when using the object, we have to de-encapsulate and re-encapsulate its properties and functions. This is done routinely, dozens of times in a work session, and very frequently at an almost subconscious level; no wonder most of my colleagues would ignore the distinction between an object and the collection of its properties as meaningless from the point of view of a working mathematician.

One of the more remarkable things about the mental objects of mathematics is that their high reliability is possible only because they can be de-assembled and assembled back in seconds, like a Kalashnikov automatic rifle, and retain their full functionality.

I came upon this Kalashnikov simile when I was a student and soon discovered that it was useful as a basis for an aesthetic judgement on films, poetry, etc. A film or a poem can be treated as a work of art only if it survives the analysis by de-assembly and re-assembly. (And it is usually a sign of a masterpiece if you have no idea whatsoever of how to de-assemble it.) Conceptual art, as a rule, fares badly in the process.

Of course, survival after a re-assembly is proof that the work of art has a meaning that is larger than the sum of its parts. In mathematics, some objects have a much greater meaning, while others are no more than convenient and disposable shorthands for the untidy sum of their parts. A new and larger meaning appears when we discover that the encapsulation leads to a wider range of interactions with other encapsulated objects. As a rule, an interaction between two capsules requires, initially, de-encapsulating them and running the interaction at a lower level. But very soon a higher-level interaction can be encapsulated on its own, which allows us to forget about the lower-level interactions. We resort to de-assembly when we lose our confidence that our higher-level work is correct.

This explains an important difference between mathematics and many other human activities: in order to secure a certain level of mathematical skills, the learner has to learn the next, higher level. Indeed, mathematical objects, concepts, and procedures are interiorized in good working condition only if they can be assembled into higher-level mathematical constructs. The ability to solve routine, rote learned problems at a certain level L is not proof that one understands mathematics at level L; but the ability to apply L level mathematics within routine problems at the next level $L + 1$ is proof that one has mastered level L. To drive a car, one does not have to be trained as a Formula 1 racer, but to teach mathematics at the high school level, a teacher has to have a knowledge of university level mathematics. The same principle applies throughout the entire range of applications of mathematics. Investment banks hire people with a PhD in mathematics or physics for jobs which require just a good knowledge of university level mathematics and statistics. This also means that the work of a mathematics teacher should be assessed not by the exam results of his students, but by their success at the next level of education. In terms of the English educational system, the success of a GCSE level mathematics teacher should be measured by the number of his/her students who take mathematics at A level and by their performance there. Similarly, the best measure of a work of an A level teacher is the number of his/her students who choose to pursue a mathematically intensive degree at a university and by their performance at the university.

In everyday mathematical practice, the chain of encapsulations is so long and is done so routinely that, for a mathematician, the question about the difference between an object and the sum of its properties and functions is more or less vacuous.

The second natural question is about the identity criteria for mental objects. Is my "2" the same as your "2"?

My answer is that identity, equality, and equivalence are mathematical, not metaphysical, concepts. Objects of mathematics do not exist on their own; therefore the answer depends on the conceptual framework (and the question is meaningless outside of the shared conceptual framework).

I once taught Peano arithmetic to my students (as a part of a proof of Gödel's Incompleteness Theorem) and greatly enjoyed explaining to them that $s(s(0))$ (the result of two consecutive applications of the successor symbol s to the symbol 0) is *not* the number 2;

it is not a number at all and should be treated as a numeral, moreover, as a specific kind of numeral, like the ones for counting sheep (see page 85). This is why sheep numerals appeared in this book: I actually used them as an example in my class. There are instances when we have to carefully treat isomorphic (or equivalent, or similar, or congruent) objects as not necessarily identical. But under different circumstances, we may flick the isomorphism/identity switch with the rapidity comparable only to the "remarkable rapidity of the motion of the wing of the hummingbird"; this happens, for example, almost every time we use the language of categories.

My knowledge of the philosophy of mathematics is limited; in what I read, one of the most disturbing findings is that philosophers tend to think mostly about monumental, seminal conceptual transformations, like, for example, the emergence of the concept of isomorphism. In the history of mathematics, some of these transformations took centuries to develop. But once established, they were eventually compressed, in everyday mathematical work, into instantaneous automatic operations, frequently performed at a semiconscious level. I have never seen a discussion of whether this truly dramatic compression changes the logical and philosophical status of the transformation.

For example, the axiomatic method, first used by Euclid in his *Elements*, for centuries remained confined to Euclidean geometry and was seen as something exceptional, axioms of geometry being entities of absolute value in themselves. Nowadays, the axiomatic method is a routine tool. Indeed, to keep your work tidy, you make the list of assumptions, call them axioms, and try to stick to them in all your deductions and calculations; when you have to consider the same problem in a different context, you have to check that the axioms remain valid; if so, you can automatically apply all your previous results. (This is the reason why *axiomatization* is included in the list of methods of reproduction.) Nowadays, axiomatic systems are cheap and disposable; more of them die in waste basket than find their way into final versions of published texts.

I argue that we cannot understand how mathematics works unless we have a close look at its disposable elements, ephemera, and mundane and minute activities. In doing so, it would be interesting to trace the difference between the *atomic*, unsimplifiable objects of mathematics and more complex constructs, which, at first glance, also behave as elementary particles but contain, in a compressed and encapsulated form, whole mathematical universes.

We cannot understand how mathematics works unless we have a close look at its disposable elements, ephemera, and mundane and minute activities.

12.2 The little green men from Mars

> *It depends upon what the meaning of the word "is" is.*
> Bill Clinton

I love the dangerous imprecision of the Davis-Hersh definition of mathematics as a study of mental objects with reproducible properties: it does not specify *who* is carrying out the study. As an apocryphal saying (allegedly taken from an undergraduate student's history essay) goes,

> all history is bias because humans are observed by other humans and not by independent observers of other species.

Therefore the old chestnut,

> would mathematics be different if it was created by little green men from Mars?,

is made redundant by the Davis-Hersh definition; the real question is,

> what would the study of *human* mental objects with reproducible properties look like if it were carried out by little green men from Mars?

Jokes aside, the Davis-Hersh definition allows us to take a detached, calm look at mathematics. For want of little green men, a mathematical model is probably the most detached way of looking at the object of study. This raises the tantalizing question about the possibility of the development of *mathematical* models of mathematical cognition in humans. I would like to believe that such models will soon become possible. On several occasions in this book we have seen that concepts and constructions of computer science provide useful metaphors for understanding mathematical thinking. Why not pursue this line further and turn computational cognitive science, an emerging and lively discipline, towards the understanding of mathematical cognition?

Indeed, metamathematics, the mathematical study of the structure and properties of mathematical theories, is a well-established area of mathematical research. The emergence of *cognitive metamathematics* would be a natural next step; it would change the landscape of the philosophy of mathematics. For me, it would make the philosophy of mathematics real fun. I am prepared to wait.

Meanwhile, I wish to paraphrase President Clinton's remark and ignore the internal contradictions and difficulties which might exist in the word "exist" as applied to mathematical objects. But, since my book touches on issues of mathematical education, I feel that I also have to demarcate the boundary between my approach to the issues of mathematical practice and the methodology of theoreticians of mathematical education.

12.3 *Better Than Life*

> *It is a mistake to identify reality with the external world only.*
> *Nothing is more real than a hallucination.*
> Leslie A. White [246, p. 306]

Which verb is given by the Shorter Oxford Dictionary *as the opposite of "to don"?*

I like Anna Sfard's metaphor of mathematics as a virtual reality game: you don [?] a helmet with visors and a glove with motor sensors and suddenly see a world where you can move objects [131]; the movements of your hand (which appear to a real life observer as erratic spasms or meaningless fidgeting) are, in your virtual world, purposeful actions. Anna Sfard calls the process of the "virtual reality" objectivization of mathematical activities *reification*; see its discussion in Section 6.1.

I have to mention, as a brief side remark, that we have to remember that in the expression "virtual reality" the key word is "reality". Virtual reality is interesting or useful mostly because it represents a real world (or, as frequently happens in computer games, some enhanced and twisted version of a real world). In one of the schools in my neighborhood, teachers turned a blind eye to schoolchildren who installed on the school's computer network a multiplayer shooting game (in return, the kids administered the network, and with reassuring competence). The teachers would probably have been less complacent if they had known that their students spent long hours importing into the game the detailed layout of the school building—for use as the game space.

Mathematics is an interactive multiplayer game. Its virtual reality is constantly affected by your actions and by the actions of other players.

Why is it one single game for the entire world and thousands, if not millions, of players?

What makes the game stable?

Why does it not crash?

What is the nature of the shared game space for all players?

To anyone who wishes to discuss the psychology of virtual reality games, I strongly recommend that he/she start by watching the cult British TV sci-fi series *Red Dwarf* and by reading the linked books by the author of the script, Grant Naylor (the title of this section is borrowed from [431]—one of the best books in the series). In one episode the heroes wake up from what they perceived was their life but was, in fact, total immersion into a virtual reality game—only to be asked, patronizingly, by an attendant of the games parlor: "So, you say that you have never even made it to the Planet of Nymphomaniacs?" I think that almost every working mathematician has had such moments of acute embarrassment in conversations with his colleagues who were aware of *his* missed opportunities in *his* virtual reality of mathematics.

Developing the "virtual reality" metaphor, we come to the conclusion that mathematics can be compared to *Massively Multi-player Online Role-Playing Games* (MMORPG), a growing phenomenon of modern culture and—which is perhaps surprising—of the modern economy. (See the book by Edward Castronova [261] who was one of the first to study virtual worlds and their economies.)

Here is a question to all philosophers of mathematics: what is the nature of the intrinsic and unintended laws of MMORPGs? Why do virtual world economies of MMORPGs obey the same laws as the real world economies? In particular, why do many virtual worlds suffer from inflation?[1]

Virtual worlds of MMORPGs are an exciting and increasingly important topic; however, the scope of this book does not allow me to venture into their macroeconomic analysis. To keep the discussion on a lighter note, I quote from a BBC online article about real-world criminals targeting the virtual economies:

My fellow mathematicians, have you ever had the feeling that your magic sword got blunted? Or even stolen?

> The police are really good at understanding someone stole my credit card and ran up a lot of money. It's a lot harder to get them to buy into someone stole my magic sword.

My fellow mathematicians, have you ever had the feeling that your magic sword got blunted? Or even stolen?

Mathematical educators and theorists of mathematical education can ignore questions about the nature of the virtual reality of mathematics. In our compartmentalized world, the maintenance of mathematics as a functioning system is not their responsibility. But I am at least aware of these thorny issues; I simply try to touch on them with caution in my book.

12.4 The vivisection of the Cheshire Cat

In this book, I ignored global vistas and, instead, looked at mathematics at the level of the individual's cognition, at the level of the individual's brain. My approach was strictly local; I looked at one proof or one example at a time, or even at one elementary step in a proof at a time. I understand, of course, that mathematical practice can be fruitfully studied from the "global" point of view (an example of such an approach can be found in Corfield [16]), but I have not attempted anything like that in this book.

As the reader has had a chance to see, the principle thesis of my book is that

we cannot understand the nature of mathematics without understanding first the interaction between *learned and/or invented* mathematical processes and underlying powerful *built-in*, inborn algorithms of our brain.

Philosophers of mathematics love to concentrate on its most abstract features, elusive like the grin on the Cheshire Cat. I propose that we vivisect the Cat.

Mathematics is already quite an abstract discipline. Not satisfied with that, philosophers of mathematics love to concentrate on its most general and therefore abstract features, elusive like the grin on the Cheshire Cat. [2] My approach is different: I propose that we vivisect the Cat. I do not claim, however, that the vivisection will instantly resolve the mystery of the famous grin; I believe only that it will usefully reformulate the problem.

The main reason why I am attracted to my thesis is that it invites a serious investigation by both mathematicians and neuroscientists.

- It appears that some modules of our cognitive system behave as if they were implementations of mathematical algorithms. For example, I have suggested, in Section 4.3, that *order* appears to be built into the mechanisms for the processing of information coming from every sensory system—from smell to vision—and our brains easily build "conversion scales", monotone maps from one ordered system into another.

  In a foreseeable future, will it become possible to detect the underlying neural activity by direct measurements—say, by PET (proton emission tomography) scans, or by more sophisticated techniques of the future?
- Will it be possible to detect experimentally a response in the innate structures of the brain (say, in the visual processing centers) to conscious mathematical thinking of the brain's owner?
- Will it be possible to differentiate between "less intuitive" and "more intuitive" mathematical concepts and processes by directly measuring the responses of the brain?
- Will it be possible to identify the *mathematical* characteristics of those structures and processes which have a natural affinity to the intrinsic mechanisms of the brain?
- Will it be possible to classify, mathematically, "atomic" mathematical structures and processes—that is, those that share characteristics of "in-built" mathematical structures and processes of our brains?

If experimental findings happen to lend support to this program of "neurological reductionism", they will trigger the need for some very serious metamathematical analysis. Indeed, it will become a serious task for mathematicians to explain the surprising unity of mathematics by showing *mathematically* how the working of "atomic" mathematical processes in individual brains leads to the reproduction of mathematical objects, structures, and algorithms shared, as a result, by almost all of the human race.

Therefore I propose that, in the context of mathematical cognition, at least some of the philosophical problems of mathematics should be supplanted by explicit metamathematical questions. My proposal, of course, will remain unsubstantiated unless I demonstrate at least one example.

A first approximation to such a question can be found in Section 12.5. My discussion there starts from a sci-fi premise of the proverbial little green men from Mars stealing a satellite from its orbit around Earth and attempting to analyze the dedicated microchip responsible for the encryption of communication channels. This sci-fi scenario is not as superficial as it seems. The standard method of experimental cognitive psychology is response time analysis: a series of tasks is given to the subject and the response time is recorded. One of the topics touched upon in this book, the difference between subitizing (immediate recognition of small numbers—two, three, or four—of objects without counting) and the proper counting of objects, one by one, was studied by response time analysis.

But the response time analysis of experimental psychology is strikingly similar to the so-called timing attacks on embedded cryptographic devices, say, microchips in credit cards, when information about the secret keys on the card is deduced by analyzing its response time [374]. Moreover, brain scan techniques where the activities of various parts of the brain are studied by measuring the blood flow

> *In the context of mathematical cognition, at least some of the philosophical problems of mathematics should be supplanted by explicit metamathematical questions.*

(which correlates with energy consumption) also have an analog in cryptanalysis—namely *power trace analysis*: the secret key on a credit card is deduced from the pattern of its energy consumption when it is plugged into the card reader [329, 375].

So, suppose little green men from Mars were to study one of the mathematical devices produced by humans. Would they be surprised to discover that humans were using particular mathematical structures? I discuss that in the Section 12.5 and argue that this is not a philosophical but a *mathematical* question.

12.5 A million dollar question

Zadaje pytania wymijajace,
by przeciac droge wymijajacym odpowiedziom.
I ask circumspect questions
to avoid circumspect answers.
Stanislaw Jerzy Lec

Whenever you discuss the realities of mathematical life, words
of wisdom from Vladimir Arnold, an outspoken critic of the *status
quo* of modern mathematics, often help to clarify the issues:

> All mathematics is divided in three parts: cryptography
> (paid for by the CIA, the KGB and the like), hydrodynam-
> ics (supported by manufacturers of atomic submarines) and
> celestial mechanics (financed by the military and by other
> institutions dealing with missiles, such as NASA).
>
> Cryptography has generated number theory, algebraic ge-
> ometry over finite fields, algebra, combinatorics and com-
> puters. [2]

Let us pause here and, ignoring the bitter irony of Arnold's words,
take them at face value. Indeed, why is there so much fuss sur-
rounding finite fields? Why is modern, computer-implemented cryptography based on finite fields?

Why does mathematics reuse, again and again, the same objects?

This question is relevant to our
discussion since computers can be
viewed as *very* crude models of a
mathematical brain. One of the major problems of human mathematics
is why mathematics chooses to op-
erate within a surprisingly limited
range of basic structures. Why does it
reuse, again and again, the same objects? It is this aspect of mathematical practice that turns many mathematicians into instinctive
(although not very committed) Platonists.

But why not ask the same question about computers? Let us
make it more specific: why is the range of structures usable in
computer-based cryptography so narrow? Unlike the philosophical
questions of mathematics, this last question has the extra bonus of
having very obvious practical implications.

Imagine that the proverbial little green men from Mars stole
a satellite from its orbit around Earth and attempted to analyze
the dedicated microchip responsible, say, for the Diffie-Hellman key
exchange. Would they be surprised to discover that humans were
using finite fields and elliptic curves?

For further discussion, we need some details of the Diffie-Hellman key exchange. I repeat its basic setup in a slightly more abstract way than is usually done.

Alice and Bob want to make a shared secret key for the encryption of their communications. Until they have done so, they can use only an open line communication, with Clare eavesdropping on their every word. How can they exchange the key in open communication so that Clare will not also get it?

The famous Diffie-Hellman key exchange protocol is a procedure which resolves this problem. Historically, it is one of the starting points of modern cryptography. In an abstract setting, it looks like this:

- Alice and Bob choose a big finite abelian group G (for our convenience, we assume that the operation in G is written multiplicatively). They also specify an element $g \in G$. (As the result of her eavesdropping, Clare also knows G and g.)
- Alice selects her *secret* integer a, computes g^a and sends the value to Bob. (Clare knows g^a, too.)
- Similarly, Bob selects his *secret* integer b and sends g^b to Alice. (Needless to say, Clare duly intercepts g^b as well.)
- In the privacy of her computer (a major and delicate assumption) Alice raises the element g^b received from Bob to her secret exponent a and computes $(g^b)^a$. (Clare does not know the result unless she has managed to determine Alice's secret exponent a from the intercepted values of g and g^a.)
- Similarly, Bob computes $(g^a)^b$.
- Since $(g^b)^a = g^{ab} = (g^a)^b$, the element g^{ab} is the secret shared element known only to Alice and Bob, but not to Clare. The string of symbols representing g^{ab} can be used by Alice and Bob as their shared secret key for encryption of all subsequent exchanges.[3]

So, what do we need for the realization of this protocol? I will outline the technical specifications only in a very crude way. They can be refined in many ways, but there is no need to do that here; a crude model will suffice.

Since we can always replace G with the subgroup generated by g, we can assume that G is cyclic. We will also assume that G has prime order p.

Therefore, to implement the Diffie-Hellman key exchange, we need the following:

- A cyclic group G of very large prime order p such that its elements can be presented by short (that is, of length $O(\log p)$) strings of 0's and 1's.
- The group operation has to be quick, in any case, better than in $O(\log^2 p)$ basic operations of the computer.

- The *discrete logarithm problem* of finding the secret exponent a from g and g^a has to be very difficult for all elements $g \neq 1$ in G; in any case, it should not allow a solution by a polynomial time algorithm.[4]
- This should preferably (but not necessarily) be done for an arbitrary prime p or for sufficiently many primes; to make the problem easier, let "sufficiently many" mean "infinitely many".
- The implementation of the particular instances of the algorithm, compilation of the actual executable file for the computer (or realization of the algorithm at the hardware level in a microchip, say, in a mobile phone), should be easy and should be done in polynomial time of small degree in $\log p$.

There are two classical ways of making cyclic groups C_p of prime order p: one of them is the additive group of the field of residues modulo p, $\mathbb{Z}/p\mathbb{Z}$. In another, we select a prime q such that p divides $q-1$ and generate G by an element g of the multiplicative order p in the multiplicative group $(\mathbb{Z}/q\mathbb{Z})^*$. In the additive group $\mathbb{Z}/p\mathbb{Z}$, the exponentiation $g \mapsto g^n$ is just multiplication by n, $g \mapsto n \cdot g$, and the Euclidean algorithm instantly solves the discrete logarithm problem. In the multiplicative group $(\mathbb{Z}/q\mathbb{Z})^*$, the discrete logarithm problem is apparently hard. It is also conjectured to be hard in the group of points of an elliptic curve over a finite field, thus giving rise to elliptic curve cryptography. Notice that, in all cases, the group, as an abstract algebraic object, is exactly the same, the cyclic group of order p; it is the underlying computational structure that matters.

How can we compare different computational structures for C_p? Look again at the examples $C_p \simeq \mathbb{Z}/p\mathbb{Z}$ and $C_p \hookrightarrow (\mathbb{Z}/q\mathbb{Z})^*$. Elements of $\mathbb{Z}/p\mathbb{Z}$ can be naturally represented as integers

$$0, 1, 2, \ldots, p-1.$$

Given an element $g \in (\mathbb{Z}/q\mathbb{Z})^*$ of multiplicative order p, we can use square-and-multiply[5] to raise g to the power of n in $O(\log n)$ time. Hence the map

$$\mathbb{Z}/p\mathbb{Z} \to (\mathbb{Z}/q\mathbb{Z})^*$$
$$n \mapsto g^n$$

gives us an isomorphism of the two computational implementations of C_p (an isomorphism which can be computed in time linear in $\log p$).[6] We shall say that the implementation of C_p as $\mathbb{Z}/p\mathbb{Z}$ is *reducible* to its implementation as $C_p \hookrightarrow (\mathbb{Z}/q\mathbb{Z})^*$. *To compute the inverse isomorphism means to solve the discrete logarithm problem*, which, as is universally believed, cannot be done in time which is polynomial in $\log p$. Therefore *morphisms* of computational structures for C_p are homomorphisms computable in polynomial time.

As Blake, Seroussi, and Smart comment in the introduction to their book on elliptic curve cryptography [312, pp. 6–8], the three types of groups we just mentioned represent the three principal classes of commutative algebraic groups over finite fields: unipotent—$\mathbb{Z}/p\mathbb{Z}$; tori—$(\mathbb{Z}/q\mathbb{Z})^*$; and abelian varieties—elliptic curves. They can all be built from finite fields, by simple constructions with fast computer implementations. So far I am aware of only one other class of computational structures for finite abelian groups proposed for use in cryptography, "ideal class groups" in number fields [324] (but it is not clear to me whether they allow a cheap mass set-up).

My million dollar question follows:

Are there polynomial time computational structures for cyclic groups of prime order (which therefore have a chance of meeting the memory and speed requirements of computer-based cryptography) which cannot be reduced, within polynomial space/time constraints, to one of the known types?

Notice that non-reducibility to $\mathbb{Z}/p\mathbb{Z}$ would mean that the discrete logarithm problem cannot be solved in polynomial time, giving such structures a chance to meet security requirements as well.

I accept that this question is likely to be out of the reach of modern mathematics. The answer will definitely involve some serious advances in complexity theory. If the answer is "yes", especially if you can invent something which is quicker than elliptic curve systems, you can patent your invention[7] and make your million dollars. If the answer is "no"—and this is what I expect—it will provide some hint as to how similar questions can be asked about mathematical algorithms and structures acceptable for the human brain and about algorithms implemented in the brain at the innate, phylogenic level.

Meanwhile, mathematics already has a number of deep results which show the very special role of finite fields in the universe of all finite structures; we shall discuss one such theorem in Section 12.6.3.

12.6 The boring, boring theory of snooks

*There are billions of gods in the world. They
swarm as thick as herring roe. Most of them are
too small to see and never get worshipped, at least
by anything bigger than bacteria, who never say
their prayers and don't demand much in the way
of miracles.*

*They are the small gods—the spirits of places
where two ant trails cross, the gods of microcli-
mates down between the grass roots. And most of
them stay that way.*

Because what they lack is belief.

Terry Pratchett, *Small Gods* [434, p. 11]

12.6.1 Why are some mathematical objects more important than others?

In Section 12.5, I tried to sketch a metamathematical question
which supplants some parts of philosophical arguments about the
nature of mathematical objects. Here is another case study in metamathematics, this time concerned not with the practical usability
of particular mathematical objects, but with their relative *importance*. I take the challenge offered by David Corfield in his beautiful
book [16].

Should we not consider it a little strange that whatever our
'ontological commitments'—a notion so central to contemporary English-language philosophy—*vis-à-vis* mathematics, they can play no role in distinguishing between entries
that receive large amounts of attention, Hopf algebras, say
[. . .], and some arbitrary cooked up algebraic entities. If I
define a *snook* to be a set with three binary, one ternary and
a couple of quartenary operations, satisfying this, that and
the other equation, I may be able to demonstrate with un-
objectionable logic that all finite snooks possess a certain
property, and then proceed to develop snook theory right
up to noetherian centralizing snook extensions. But, unless
I am extraordinarily fortunate and find powerful links to
other areas of mathematics, mathematicians will not think
my work worth a jot. By contrast, my articles may well be
in demand if I contribute to the understanding of Hopf algebras, perhaps via noetherian centralizing Hopf algebra
extensions. [16, p. 11]

Why are some mathematical objects worshipped while others are ignored? I am in complete agreement with David Corfield that the "ontological commitment", acceptance of some form of existence of mathematical objects in some form of "ideal" reality, does not help to explain the

Even if mathematical objects exist in the strictest Platonic sense, almost all of them live out their miserable existence in complete obscurity.

striking disparity in importance among possible mathematical theories. Do I need to remind the reader that the potential multitude of mathematical objects is infinite? Even if mathematical objects exist in the strictest Platonic sense, almost all of them, like the "small gods" of Terry Pratchett, live out their miserable existence in complete obscurity.

When you talk to mathematicians, you easily discover that, according to widely held opinion, there are basic reasons why particular mathematical objects, structures, or theories deserve to be studied:

A. they have applications, or claim to have applications, outside mathematics, say, in physics or communications technology;
B. they are intrinsically beautiful and allow for a nice theory with elegant proofs;
C. they have strong internal connections to other interesting objects of mathematics;
D. they appear as examples or can be applied to the solution of problems arising in theories which are interesting for reasons A, B, or C; or
E. they serve as a metatheory explaining the nature and structure of other interesting theories.

In a classical textbook on abstract algebra, I. N. Herstein [300] emphasizes the special role of reasons D and E:

> In abstract algebra we have certain basic systems which, in the history and development of mathematics, have achieved positions of paramount importance. These are usually sets on whose elements we can operate algebraically—by this we mean that we can combine two elements of the set, perhaps in several ways, to obtain a third element of the set— and, in addition, we assume that these algebraic operations are subject to certain rules, which are explicitly spelled out in what we call the axioms or postulates defining the system. In this abstract setting we then attempt to prove theorems about these very general structures, always hoping that when these results are applied to a particular, concrete realization of the abstract system there will flow out facts and insights into the example at hand which would have

been obscured from us by the mass of inessential information available to us in the particular, special case.

We should like to stress that these algebraic systems and the axioms which define them must have a certain naturality about them. They must come from the experience of looking at many examples; they should be rich in meaningful results. One does not just sit down, list a few axioms, and then proceed to study the system so described. This, admittedly, is done by some, but most mathematicians would dismiss these attempts as poor mathematics. The systems chosen for study are chosen because particular cases of these structures have appeared time and time again, because someone finally noted that these special cases were indeed special instances of a general phenomenon, because one notices analogies between the highly disparate mathematical objects and so is led to a search for the root of these analogies.[8]

Almost all randomly defined algebraic structures do not allow for any kind of theory of any structural interest.

Of course, an interesting theory might belong to several types simultaneously. Reason B for being interesting presents most of the difficulties when you try to understand the prominence of particular objects of theories. If a theory or an object is selected for development or study solely for reasons of its elegance and beauty, why can we not pick instead some other theory, with equally beautiful statements and elegant proofs?

My answer is: because they are exceptionally rare.

12.6.2 Are there many finite snooks around?

To be more concrete, I wish to turn to David Corfield's example of

"snooks ... satisfying this, that and the other equation",

and formulate the following metamathematical thesis:

Almost all randomly defined snooks and, more generally, almost all randomly defined algebraic structures, do not allow for any kind of theory of any structural interest.

In this form, my thesis is still excessively general; with respect to snooks, proposed by David Corfield, it can be specialized and turned into an explicit and provable (or refutable) metamathematical conjecture.

Indeed, let us denote the five operations on snooks by the functional symbols

$$A(\cdot,\cdot), B(\cdot,\cdot), C(\cdot,\cdot), T(\cdot,\cdot,\cdot), Q(\cdot,\cdot,\cdot,\cdot).$$

We define, as is done in universal algebra, a *variety* of snooks as the set of all snooks which satisfy some fixed *identities*, that is, equalities of the kind, say,

$$Q(T(x,y,z), A(x,y), B(y,z), C(z, T(x,y,u))) = A(B(x,y), C(x,y))$$

which hold for all values of variables involved. Here I wrote a random identity, caring only about the number of variables involved in every function being in agreement with the assumption that A is a binary operation, etc. Different varieties of snooks are defined by different sets of identities, while the same variety can be defined by many different (but equivalent) sets of identities. We shall look only at those varieties which are defined by finitely many identities. Then, with every variety S of snooks, we can associate its *definition length* $dl(S)$, the minimal possible total length of identities which defines S.

Simon Thomas,
aged 2

A snook is called *constant* if every function A, B, C, T, Q on it is constant, that is, takes just one value. Obviously, a snook with just one element (we call it *trivial*) is constant, and every variety of snooks contains the trivial snook.

THE RANDOM SNOOK CONJECTURE, I. There is a constant $c > 0$ such that, given a variety S of snooks chosen at random from all varieties of snooks of definition length at most n, the probability that S contains a non-constant finite snook is smaller than e^{-cn}.

In less formal terms, it means that the probability for a random variety of snooks to contain a non-constant finite snook decreases exponentially fast when the definition length grows.

In a similar vein, it is possible to formulate other conjectures, this time turning our attention to individual snooks. For example, within each variety S one can consider *finitely presented* snooks, that is, snooks given by a finite number of generators, say, $a, b, c, \ldots$, and by finitely many relations, that is, equalities of the kind

$$A(a,b) = T(a,a,b),$$

etc.

THE RANDOM SNOOK CONJECTURE, II. With probability which tends to 1 exponentially fast with the growth of the total length of defining identities and relations, a random finitely presented snook is infinite and does not allow non-trivial automorphisms.

Formulating conjectures about random snooks very soon becomes no less boring than studying them. The common point of all my conjectures is simply that random snooks are boring, there

is nothing of substance to study in them, they do not allow a sufficiently rich structured theory of any kind; even the formulation of
a readable result is virtually impossible.

I stress that all my conjectures are mathematical statements
which are likely to be provable or refutable (I would be extremely
surprised if the Random Snook Conjectures are independent of the
Zermelo-Fraenkel axioms of set theory).

My confidence stems from the first results on "random groups"
in group theory, theorems like Olshanski's [393], which claims that
a random finitely presented group is hyperbolic and hence allows
a uniform approach to proving everything which can be said about
it, by industrial strength methods from the theory of hyperbolic
groups. Even more telling is a theorem by Kapovich and Schupp
[370] which says that, for random one-relator groups, every classical problem of group theory has a more or less trivial solution. Random objects bear no distinctive features; within the class of groups
(one of the nicer objects of mathematics) this means that their properties are nice and easy. In a random class of snooks, or some other
random variety, there is no reason to expect random snooks to behave nicely. Moreover, random snooks are far from *unique* in their
boringness, and for that reason we are unlikely to be able to prove
anything interesting about random snooks.

12.6.3 Snooks, snowflakes, Kepler, and Pálfy

By varying the defining identities for snooks, we can get an infinite
multitude of algebraic structures, each different from the others
and each having the same right to exist. This is a classical example of a "bad", unstructured, uncontrolled infinity. When encountering such situations, mathematicians professionally try to introduce
some structure into the disorder and to find general principles gov-
erning the universe of snooks.

The paradigm for such an approach is set in Kepler's classical
work on snowflakes [55]. Of the myriads of snowflakes, there are
no two of the same shape; however, almost all of them exhibit the
strikingly precise sixfold symmetry. Kepler's explanation is breath-
takingly bold: the symmetry of snowflakes reflects the sixfold symmetry of the packing of tiny particles of ice (what we would now
call atoms or molecules) from which the snowflake is composed.
In 1611, when the book was written, it was more than a scientific
conjecture—it was a prophecy.

Of the infinitely many possible algebraic laws defining generalized snooks, some may allow for the existence of a finite structure.
I will now outline a "snowflake" theory of arbitrary finite algebras,[9]
which will of course cover the case of finite snooks. The theory belongs to David Hobby and Ralph McKenzie [365]; to avoid excessive

Fig. 12.1. Classical photographs of snowflakes by Wilson Bentley, c. 1902.
Source: *Wkipedia Commons*. Public Domain.

detail, I will concentrate on its key ingredient, a theorem by Péter
Pál Pálfy [394] on the structure of "minimal" algebras ([365, Theorem 4.7]).

The key idea is that we study finite algebras up to *polynomial equivalence*: we associate with every algebra $\mathbb{A}$ with
ground set A the set of *all polynomial functions* on A, that is,
all functions from A to A expressible by combination of basic algebraic operations of $\mathbb{A}$, with arbitrary elements from
A being allowed to be used as constant "coefficients". For example, if $\mathbb{S}$ is a finite snook with the set S of elements and s
is a fixed element of S, then

$$A(s, x)$$

is a polynomial function of a single variable x, while

$$T(x, A(s, y), x)$$

Péter Pál Pálfy,
aged 13

is a polynomial function of two variables x and y. Two algebras are said to be *polynomially equivalent* if they have the same ground set and the same sets of polynomial functions. In particular, this means that every basic algebraic operation of the first algebra is expressed in terms of the operations of the second algebra, and vice versa. If we ignore the computational complexity of these expressions (which is not always possible in problems of a practical nature; see Section 12.5), the two algebras are in a sense mutually interchangeable.

Given a finite algebra $\mathbb{A}$, a polynomial function $f(x)$ in a single variable induces a map from A to A. Since A is finite, either $f(x)$ is a permutation of A, or it maps A to a strictly smaller subset $B \subset A$. In the second case, some iteration

$$g(x) = f(f(\cdots f(x) \cdots))$$

is an idempotent map:

$$g(g(x)) = g(x)$$

for all x. The idempotency of g allows us to "deform" and squeeze the basic operations of $\mathbb{A}$ to the set $C = g[A]$. If, for example, $T(\cdot, \cdot, \cdot)$ were an operation of $\mathbb{A}$, $T' = g(T(\cdot, \cdot, \cdot))$ becomes an operation on C. Adding all polynomial operations of $\mathbb{A}$ which preserve C, we get a new algebra $\mathbb{C}$ (we shall call it a *retract* of $\mathbb{A}$) which carries in itself a considerable amount of information about $\mathbb{A}$. For example, every homomorphic image of $\mathbb{C}$ is a retract of a homomorphic image of $\mathbb{A}$ [395].

But what happens if a finite algebra $\mathbb{A}$ has no proper retracts (that is, with C being a proper subset of A) and is therefore unsimplifiable? Pálfy calls such algebras *permutational*. Assuming that the algebra has at least three elements[10], we have a further division:

1. Every polynomial function defined in terms of $\mathbb{A}$ effectively depends on just one variable. Then all polynomial functions on A are permutations, and $\mathbb{A}$ is polynomially equivalent to a set A with an action of a finite group G, where the action of each element $g \in G$ is treated as a unary operation. This case is not at all surprising.
2. In the remaining case, when $\mathbb{A}$ is sufficiently rich for the presence of polynomial functions which really depend on at least two variables, the result is astonishing: $\mathbb{A}$ *is polynomially equivalent to a vector space over a finite field!*

So finite fields appear to be more important, or more basic, than finite snooks. Pálfy's theorem is a partial explanation of the mystery which we have already discussed in this chapter: *why are finite fields so special?* Mathematics needs more results of this nature, which help to clarify and explain the hierarchy of mathematical objects. Without a rigorous metamathematical study of the relations between various classes of mathematical objects and without the understanding of the reasons why some mathematical structures have richer theories than other structures, it is too easy to exaggerate the role of history and fashion in shaping mathematics as we know it now. I do not believe that the ideas of social constructivism can be really fruitful in the philosophy of mathematics. However, I have no space in this book to get into a detailed discussion.

12.6.4 Hopf algebras

> *All animals are equal,*
> *but some animals are more equal than others.*
> George Orwell

Interesting objects in mathematics are rare; even the briefest look at mathematics reveals that the same objects and structures are recycled again and again. Remarkably, most of these ubiquitous structures come from physics (interpreted broadly). It is possible to suggest an explanation as to why most *rich* structures come from physics; they are structures that were selected, from the immense variety of possible mathematical structures, as models for some aspects of the physical universe. We know, from experience, that the universe is rich, diverse—but ruled by unified laws. The mathematics which models these laws must provide for a rich theory.

The case of Hopf algebras is really illuminating. It is hard to invent a mathematical object which is more closely linked to the objective reality of physics, or, more precisely, to the way in which physics describes the real world. The very fact that physics exists and is successful in its description of the world provides a strong hint that the theory of Hopf algebras can be developed in some rich detail.

Basically, physicists measure things and their states, assigning to them numeric values. Hopf algebras (and their glamorous descendants, quantum groups) come from the measurement of processes which have a group structure. This is not an excessive requirement; it simply means that the processes are invertible and their composition satisfies the associative law, the latter frequently being a consequence of the most natural assumption of the homogeneity of time.

Look, for example, at the motion of a solid body about a fixed point. This is a physical process; but rotations can be composed with other rotations, thus giving to the set of rotations the structure of a group.

Meanwhile, measurements are expensive and physicists measure one function at a time. In the case of the rotations of the solid body, we can associate with the initial position of the body some (Cartesian) coordinate vectors $\vec{e}_1, \vec{e}_2, \vec{e}_3$ and make another set of coordinate vectors, $\vec{f}_1, \vec{f}_2, \vec{f}_3$, to mark the new position of the body. Measurements, especially in quantum mechanics, are frequently formalized as projections of a vector onto a one-dimensional subspace spanned by a given vector $\vec{e}$.[11] In our particular case, the most natural quantity to measure is the length of the projection of $\vec{f}_i$ onto the direction of $\vec{e}_j$, which, of course, gives us the matrix element of the rotation matrix

$$r_{ij} = (\vec{f}_i, \vec{e}_j).$$

Therefore what we measure, when we measure rotations, are matrix elements of the rotation matrix, or, if we run a really technically intricate experiment, some more complicated functions of matrix elements.

Now our rotation R can be decomposed as a product of a pair of rotations

$$R = S \cdot T$$

in infinitely many ways. Again, these new rotations are given to us as assemblies of measurements, s_{ij} and t_{ij}. If G is the group of rotations, then matrix elements r_{ij}, s_{ij}, t_{ij} are functions on G. The composition of rotations is the map

$$G \times G \longrightarrow G;$$

hence r_{ij} can be lifted from G to $G \times G$ and can be interpreted as a function on $G \times G$. The problem of describing the group law on G in terms of measurements becomes the task of expressing the function r_{ij} as a function of two variables (S, T). But all we know about S and T is given to us as our measurements s_{ij} and t_{ij}. Luckily, we can complete the task and assign to r_{ij} its expression in terms of s_{ij} and t_{ij}:

$$r_{ik} \mapsto \sum s_{ij} t_{jk}.$$

Let me emphasize that the arrow goes in the direction opposite to multiplication (or composition) of rotations: we take a function on G and assign to it a function on $G \times G$.

The result is what we call a *Hopf algebra*. The formal definition is just a technical refinement of this simple idea. The map of the function spaces that we just constructed is called *comultiplication*; if H is the space of functions on G, then it becomes an embedding

$$H \longrightarrow H \otimes H$$

where $\otimes$ denotes the tensor product of vector spaces. I shall skip further details but note only that we also need the map

$$H \longrightarrow H$$

obtained, in a similar way, from the inversion,

$$R \mapsto R^{-1},$$

as well as the usual multiplication of functions.

It may be a gross oversimplification, but a Hopf algebra in a sense is nothing more than glorified matrix multiplication.

Why are Hopf algebras so important in quantum physics (and why is a further specialization of the concept, *quantum groups*, so fashionable in modern mathematics)? My answer is, again, a gross oversimplification, but it allows us to explain, in a few words, their exceptional role.

In classical physics, we can pretend that we have made every possible measurement and work with the collection of measurements as a single object (in our example—a rotation matrix). The characteristic feature of quantum physics is that the Indeterminacy Principle forbids measuring everything at once. At a more basic level, the set-up of quantum physics forces us to be attentive to individual functions. David Corfield rightly stresses [16, p. 24] that this leads to an emphasis on the duality between the observables and states; we not only use observables (measurements, in the jargon of the previous paragraphs) to study states, but we also use states to distinguished between observables. The resulting mathematical abstraction is something which can be explained at the level of the notational conventions of elementary high school calculus: instead of writing the value of the function f at the point x as $f(x)$, we can use a slightly stranger notation $\langle f, x \rangle$ (or even $\langle f \| x \rangle$ as physicists do) to emphasize that this is also the evaluation (or testing) of the function *by the argument x*.[12]

Hopf algebras are really basic objects of theoretical physics. But do they deserve to be one of the "simple things" of the present book? Probably not, because they are deeply counterintuitive: in our perception of the world, our sensor system measures the intensity of individual stimuli, but the processing of this information is hidden deep into the subconsciousness; the image of the world as given to us by our senses is an integral and highly distilled construct. The cognitive analogue of a physical measurement is something which happens at the level of a single receptor and a single neuron—it is beyond our conscience and our control.

12.6.5 Back to ontological commitment

We have to do mathematics using the brain
which evolved 30,000 years ago
for survival in the African savanna.
Stanislav Dehaene

Poor snooks cannot compete with Hopf algebras for the attention of mathematicians. The explanation for this disparity, as we have seen, is twofold. Most snooks are deeply irrelevant—and this fact could be and should be explained metamathematically. Mathematics, as a cultural entity, is built upon the presumption that its objects are boring unless proven interesting (I am not well read in the philosophy of mathematics and, perhaps, for that reason I have never seen this general methodological principle clearly stated in writing). Even if some snooks give rise to a degree of reasonable theory, mathematicians, for the sake of the health and sanity of their discipline, have every right to demand an advance justification of the snooks' purpose. The strongest possible justification of the purposefulness of a mathematical object comes, as in the case of Hopf algebras, from the needs of physics.

Therefore I am not in complete agreement with David Corfield when he says:

> We may have been led to use specific Hopf algebras to allow us to perform calculations with Feynman diagrams [. . .], but it cannot be right to say that they are structures instantiated in the world. Still we cannot distinguish between snooks and Hopf algebras. [16, p. 12]

I claim that we *can* distinguish between snooks and Hopf algebras. I do not know whether I agree or disagree with Corfield that Hopf algebras are not instantiated in the world; I have no firm opinion on that issue. What really matters to me is that Hopf algebras are instantiated (or at least deeply rooted) in the human practice of measuring this world.

Corfield's book made me realize that I cannot easily change my spots and get over my Vygotskian upbringing. However, David Corfield immediately deflated the triumph of my naive Vygotskianism by pointing out that Hopf algebras were invented in algebraic topology before they came into use in physics. It appears that it does not matter what we measure—physical or ideal objects! So we return to the same disturbing question about the nature of ideal objects in mathematics. I do not see an obvious way to resolve this issue.

Instead, I wish to offer to philosophers a different problem. As I am trying to demonstrate in this book, some of the simplest mathematical structures and processes have a special ontological status:

those already hardwired into our brains—such as order, symmetry, parsing rules. I believe that these atomic particles of mathematics deserve some special attention from philosophers, and I am happy to leave to them the discussion of finer details.

My own approach to the philosophy of mathematics is purely pragmatic: for me, philosophy is a useful tool for making and assessing *value judgements*, on the scale of interesting–uninteresting, important–unimportant. This places the "simple things" of this book in a

Mathematics is built upon the presumption that its objects are boring unless proven interesting.

very special position. Indeed, if we accept the special status of hardwired mathematical structures and their immediate recombinations (like the palindrome, a simplest recombination of order, symmetry, and parsing), then we have to accept that these "simple things" of mathematics are outside of the area of the applicability of value judgements because they are just part of us as human beings.

Regarding palindromes, for example, no matter what you think about their importance or irrelevance, they have existed for centuries. The fact that they can be treated as mathematical objects went more or less unnoticed. But they can; as I have shown in Section 3.4, palindromes can be used, for example, as the basis of the theory of Coxeter groups. This does not make them more important or interesting; as I said, they are *outside* of the area of value judgements.

Of course, it still gives us the right to ask technical questions about the *exact nature* of the *optimality* of brain algorithms. Given our "hardware limitations", does the saccadic movements algorithm indeed provide an optimal sampling method (see page 28)? In the literature, you can find a heated debate on whether the human eye, as a hardware device, has an optimal design and whether octopi have better eyes (see, for example, [188]). So what? We can study our eyes, but, for the foreseeable future, we have to live with the eyes we have. And we still have to do mathematics with our imperfect brains which evolved for completely different purposes.

12.7 Zilber's Field

Many mathematical objects are exceptionally rigid in the sense that we cannot change them at will: they offer resistance. The "robustness" of a mathematical object (one may also talk about a dual concept: the "robustness" of a mathematical theory which describes the object) has been explicated as a formal mathematical notion, *categoricity*, in the mathematical discipline called model theory. One of the recent results in that theory, Boris Zilber's work on the Schanuel Conjecture, can be described (but perhaps not formulated in full detail) at a level of elementary algebra/calculus/number theory. It poses a significant philosophical challenge which, to the best of our knowledge, so far has been entirely ignored by the philosophers of mathematics.

The Schanuel Conjecture is about transcendental numbers. It says that if you have n complex numbers $x_1, \ldots, x_n$ which are linearly independent over the rationals and consider the system of numbers

$$x_1, \ldots, x_n, \exp(x_1), \ldots, \exp(x_n),$$

then the latter has transcendence degree at least n. The conjecture contains in itself a huge number of known results. For example, if one takes $x_1 = \ln 2$, the conjecture says that the system

$$\{\ln 2, \exp(\ln 2)\} = \{\ln 2, 2\}$$

has transcendence degree at least one, which means that $\ln 2$ is a transcendental number—a classical result of transcendental number theory.

Zilber [418] took a number of natural (and known) algebraic properties of complex numbers and the exponentiation function $\exp$ for axioms (so that "exponentiation" is understood as a map which satisfies

$$\exp(a + b) = \exp(a)\exp(b)),$$

added to them, as a further axiom, the formulation of the Schanuel Conjecture (still unknown), and proved:

1. The axioms are consistent, that is, they have a model, an algebraically closed field $\mathbb{B}$ of characteristic 0 with a formal exponentiation function $\exp$, such that all these axioms are satisfied in $\mathbb{B}$.
2. All axioms, with a possible exception of the Schanuel Conjecture, are satisfied in the field $\mathbb{C}$ of complex numbers with the standard exponentiation.
3. There is exactly one, up to isomorphism, such field $\mathbb{B}$ of cardinality continuum.

And now we can formulate some questions:

- Why would almost every mathematician (with the possible exception of intuitionists and ultrafinitists like Yessenin-Volpin— but they have an honorable excuse) immediately agree that of course it should be true that $\mathbb{B} = \mathbb{C}$ and that therefore the Schanuel Conjecture should be true? What is the basis of this belief in "it should be true"?

- Why does Zilber's result have a suspiciously foundational, meta-mathematical feel about it?
- Zilber's field $\mathbb{B}$ had been built by a version of the Fraïssé-Hrushovski amalgam method, non-constructive and seriously transcendental. Why are most mathematicians prepared to believe that $\mathbb{B} = \mathbb{C}$, despite the two objects having completely different origins?

The appearance of the Fraïssé amalgam method on the scene should remind us that even in the countable domain we have a wonderful and paradoxical example of seemingly incompatible constructions leading to the same "universal" object.

The famous *random graph* [326] can be constructed probabilistically by coin tossing: vertices of the graph are natural numbers, and for every pair of vertices $m < n$ we toss a fair coin and, if we get heads, we connect the vertices m and n by an edge. The same graph can be constructed by a totally deterministic procedure: we take for the set of vertices the set of all prime numbers p congruent to 1 mod 4 and draw an edge between prime numbers p and q if p is a quadratic residue modulo q. In all cases the resulting object is the same, THE random graph. And—last but not least—the random graph can be constructed as a Fraïssé amalgam of finite graphs (which actually explains the first two constructions).

Is the acceptance of actual infinity just the price that mathematicians are prepared to pay for the convenience (and beauty) of the "canonical" objects of mathematics?

12.8 Explication of (in)explicitness

I have already quoted Timothy Gowers saying:

The following informal concepts of mathematical practice cry out to be explicated: beautiful, natural, deep, trivial, "right", difficult, genuinely, explanatory ...

These words deserve to be called *Gowers's program*. He formulated his program in his talk at the conference *Mathematical Knowledge* in Cambridge in June 2004.

This section will demonstrate an extreme case among many possible approaches to the explication of informal mathematical concepts: a hardcore mathematical treatment of the very concept of "explicitness".

It can be formulated as a remarkably compact thesis:

EXPLICIT = BOREL.

This thesis is promoted—perhaps in less explicit form—by Alexander Kechris and Greg Hjorth (see his survey [364]). A wonderful illustration and a template for the use of the thesis can be found in a recent work by Simon Thomas.

Simon Thomas looked at the following problem:

Does there exist an explicit choice of generators for each finitely generated group such that isomorphic groups are assigned isomorphic Cayley graphs?

Recall that if G is a finitely generated group and S is its finite generating set not containing the identity element, then the Cayley graph of G with respect to S is the graph with vertex set G and edge set

$$E = \{(x, y) \mid y = xs \text{ for some } s \in S \text{ or } S^{-1}\}.$$

For example, when $G = \mathbb{Z}$ is the additive group of integers and $S = \{1\}$ consists of the most canonical generator of integers, the number 1, then the corresponding Cayley graph is

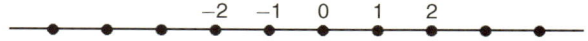

However, when $G = \mathbb{Z}$ and $S = \{2, 3\}$, then the corresponding Cayley graph is

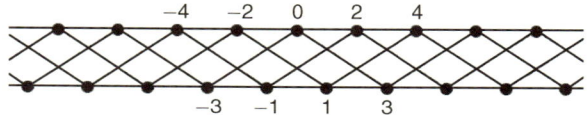

Simon Thomas's problem is very natural: is their an explicitly given *canonical* Cayley graph for each finitely generated group? The problem is about the relation between two basic concepts of algebra: that of a group and its Cayley graph; it is formulated in very elementary terms.

Simon Thomas gave a negative answer to the problem [410]. More precisely, he proved that this assignment cannot be done by a map with the Borel graph.

The underlying concepts of his proof are not that difficult. I give here only a crude description; all details can be found in Thomas's paper.

First we note that a structure of a group on the set $\mathbb{N}$ of natural numbers is given by its graph of multiplication, that is, a subset of the countable set $\mathbb{N}^3$. This subset is encoded as a (countable) sequence of 0's and 1's, hence can be viewed as a point in $2^{\mathbb{N}^3}$, the latter being equipped with the product topology. It can be shown that the space $\mathcal{G}$ of finitely generated subgroups becomes a Borel subset of $2^{\mathbb{N}^3}$, hence a standard Borel space, that is, a complete separable metric space equipped with its σ-algebra of Borel subsets (recall that a Borel algebra of a topological space is the σ-algebra generated by its open subsets; elements of a Borel algebra are called Borel sets).

At this point one needs to recall Kuratowski's Theorem:

Any standard Borel space is isomorphic to one of

(1) $\mathbb{R}$,

(2) $\mathbb{N}$, or

(3) a finite space.

In particular, the set $\mathrm{Fin}(\mathbb{N})$ of finite subsets of $\mathbb{N}$ is also a standard Borel space. Further, if X and Y are standard Borel spaces, the map

$$f : X \to Y$$

is *Borel* if its graph is a Borel subset of the direct product $X \times Y$.

Now I can state Simon Thomas's Theorem:

There does not exist a Borel map

$$f : \mathcal{G} \to \mathrm{Fin}(\mathbb{N})$$

such that for each group $G \in \mathcal{G}$:

- $f(G)$ generates G.
- If G is isomorphic to H, then the Cayley graph of G with respect to $f(G)$ is isomorphic to the Cayley graph of H with respect to $f(H)$.

Is the EXPLICIT = BOREL thesis reasonable?

Indeed, a Borel subset of $\mathbb{R}$ is any set which can be obtained from the open intervals (a, b) by performing the operations of taking countable unions, countable intersections, and complements. Well, this is a fairly wide class of sets; perhaps not everyone would agree that "Borel" is "explicit", but it is easier to accept that every "explicit" construction in the real domain produces a Borel set. Being very wide and encompassing, this explication of "explicitness" is useful for proving negative results, when we want to show that some construction cannot be made explicit.

Notice also that it is Kuratowski's Theorem that brings the feeling of precision and universality into the EXPLICIT = BOREL thesis; without it, the thesis would be much more vague.

The EXPLICIT = BOREL thesis for the real domain is an example of an "under the microscope" approach to mathematics; it is like cytology, which treats living tissue as an ensemble of cells. Representing everything by 0's and 1's gives a very low-level look at mathematics; it is fascinating that this approach leads to explicit "global" results.

The concept of a Borel set belongs to the area of mathematics called descriptive set theory. I have special feelings for descriptive set theory and am happy to see its remarkable revival. It so happened that, in my personal mathematical education, my first ever serious course (during the penultimate year at Fizmatshkola, the preparatory boarding school of Novosibirsk University) happened to be on descriptive set theory. Our lecturer based the course on the classical (and pretty archaic in its language) memoir by René-Louis Baire on pointwise limits of continuous functions [310]. I always thought that descriptive set theory was fun, but it was Thomas's Theorem that opened my eyes to its impressive explanatory power which extends well beyond analysis.

12.9 Testing times

Mathematics is part of Physics.
Physics is an experimental discipline, one of the natural sciences.
Mathematics is the part of Physics where experiments are cheap.
V. I. Arnold [104]

While writing this book, I gradually found myself busy inventing schemes of psychological experiments (preferably accompanied by brain scans) which would, I believe, shed some light on what is happening in the basic processes of the interiorization of mathematics.

I always thought that mathematics was difficult. But when I started to think about the practical side—and cost—of setting up experiments, say, concerned with the perception of symmetry by blind people, I got pretty scared. And this is only one of many experiments which might perhaps be relevant to understanding mathematical cognition. I realized that I have a great respect for experimental psychologists and neurophysiologists, and I would really like to hear their opinions from them on the issues raised in this text.

If we want to understand the relations between the numerosity of collections of discrete objects and the magnitude of continuous parameters (length, for example), it is worth setting up some experiments where the response to numerosity is analogue and continuous.

Despite all the differences in research cultures, we mathematicians have to start a dialogue with cognitive scientists and neurophysiologists. There is too much at stake here for the future of mathematics as a discipline.

I wrote this book with the hope that some of its ideas can be converted into a feasibility study for a more systematic research program into the cognitive mechanisms of mathematical practice. The program, of course, should start with a comprehensive literature search. It is likely to be an unusually difficult exercise because the questions we (mathematicians) tend to ask are likely to be subtly different from the ones which were asked by psychologists and neurophysiologists.

I wish to offer an example of such a question; I raised it at a *Mathematical Knowledge* conference in Cambridge in June 2004. Judging from the experts' responses, the question apparently has not been asked before and has not been checked experimentally.

In experimental studies of the differences between exact and approximate arithmetic, the subjects are usually asked to estimate the number of items in a group or to compare two groups by size. It is easy to notice that the responses are usually verbal (numeral) or discrete (pointing to a bigger group). If we want to understand the relations between the *numerosity* of collections of discrete objects and the *magnitude* of continuous parameters (length, for example), it is worth setting up some experiments where the response to numerosity is analogue and continuous. For example, the subjects could be instructed to pull the lever with different force according to the number of items shown. Will the subitizing/counting threshold manifest itself with the same prominence as in the case of verbal or symbolic responses?

Any serious research program in mathematical cognition should start with scanning the huge body of psychological and neurological literature for properly documented case studies of various neurological and psychological conditions, including those not having, at first glance, any relation to mathematics. On a personal level, I would like to read more about the psychological peculiarities which I occasionally observed in my students and colleagues. I give one example here: I knew a schoolboy who was able to solve very tricky non-standard plane geometry problems but whose mind went blank on problems in three-dimensional geometry. He also suffered from an extreme form of vertigo: even an attempt to stand up on tiptoes inevitably led him to a panic attack. Was that all emotional? Are there deeper links between the defects of spatial perception and vertigo? Were there studies which confirmed and analyzed—or disproved—such links?

Of course, the history of mathematics, especially of the early stages of its emergence and development, is also highly relevant to any serious look at mathematical cognition. Also, it is hard to overestimate the importance of the huge body of knowledge accumulated by anthropologists, ethnographers, and scholars of "ethnomathematics". The study of mathematical cognition is a vast, challenging, and timely task. I have written this book in the hope that some of the readers will contribute (or are already contributing) to the development of this new discipline.

Notes

[1]VIRTUAL ECONOMIES. I am not in a position to make any judgment on MMORPGs. I follow advice from an expert, Grax. He wrote to me:

> My economist friend and occasional virtual world buddy recommends looking at EVE Online's inflationary economy and comparing it to the non-inflationary economies of *Second Life* and *Project Entropia*, whose currencies are tied to the dollar. [...] I can add that each expansion in a virtual world causes a spike in inflation, since new lands are populated with significantly superior items, etc., to give the players an (extra) incentive to purchase new content.

[2]Recall the classical passage from *Alice in Wonderland*:

"I wish you wouldn't keep appearing and vanishing so suddenly; you make one quite giddy!"

"All right," said the Cat; and this time it vanished quite slowly, beginning with the end of the tail, and ending with the grin, which remained some time after the rest of it had gone.

"Well! I've often seen a cat without a grin," thought Alice; "but a grin without a cat! It's the most curious thing I ever saw in all my life!"

[3]Gregory Cherlin commented that the following is likely to be a general principle: shared knowledge is a form of commutativity.

[4]Even if a polynomial time algorithm is practically unfeasible, its very existence will undermine the commercial confidence in the cryptographic product since it potentially opens up a venue for possible improvements which will eventually destroy the cryptosystem. Commercial and military users of cryptographic products are not willing to take such risks.

[5]The following example shows how we can compute 3^{100} mod 101 using only 8 multiplications, and not 100 as one might think.

First write 100 as the sum of powers of 2:

$$100 = 64 + 32 + 4 = 2^6 + 2^5 + 2^2.$$

Then we have

$$\begin{aligned}
3^{100} &= 3^{2^6 + 2^5 + 2^2} \\
&= 3^{2^6} \cdot 3^{2^5} \cdot 3^{2^2} \\
&= \left(\left(\left(\left(\left((3^2)^2\right)^2\right)^2\right)^2\right)^2\right)^2 \cdot \left(\left(\left((3^2)^2\right)^2\right)^2\right)^2 \cdot (3^2)^2 \\
&\qquad \left[\text{ since } (3^2)^2 = 81 \equiv -20 \text{ mod } 101 \right. \\
&\qquad\quad \left. \text{and } \left(\left((3^2)^2\right)\right)^2 \equiv 400 \equiv -4 \text{ mod } 101\right] \\
&= \left(\left((-4)^2\right)^2\right)^2 \cdot \left((-4)^2\right)^2 \cdot 81 \text{ mod } 101 \\
&= 256^2 \cdot 256 \cdot 81 \text{ mod } 101 \\
&= 54^2 \cdot 54 \cdot 81 \text{ mod } 101 \\
&= 1 \text{ mod } 101.
\end{aligned}$$

[6]An impressive body of mathematics has been developed over the last half century for efficient implementation of exponentiation, with the likes of Paul Erdős and Donald Knuth involved; see [372, Section 4.6.3].

[7]I have seen a patent application for the use of the formula $x^{a+b} = x^a \cdot x^b$ in cryptography.

[8]"TOPICS IN ALGEBRA" BY I. N. HERSTEIN. I love *Topics in Algebra* [300] because it was my first textbook of abstract algebra. I read it when I was a student at Fizmatshkola, preparatory boarding school at Novosibirsk University. The book (in English) was borrowed from the library of the Sobolev Institute of Mathematics. In the boarding school, on Thursdays we were free from classes, and we were supposed to spend the day in

independent studies. I spent the day lying in bed and reading, in arbitrary order, *Topics in Algebra* by Herstein, *Finite Groups* by Daniel Gorenstein [359], and Robert Louis Stevenson's stories. A novice in English, I loved R. L. Stevenson for his exceptional clarity of language.

[9]The word "algebra" is understood here in terms of the "universal algebra" as a set with some operations of an arbitrary nature; our snooks are algebras.

[10]ONE-ELEMENT ALGEBRAS are of no interest, while every two-element algebra can be polynomially expressed in terms of the Boolean algebra

$$\langle \{\mathbf{0}, \mathbf{1}\}; \vee, \wedge, \neg \rangle$$

by a theorem about the disjunctive normal form for a Boolean function. Up to polynomial equivalence there are six (or—if we ignore the permutation of the base set—seven) two-element algebras:

- no operation;
- negation $\neg$;
- "plus" $+$ (the cyclic group of order 2);
- "and" $\wedge$;
- "or" $\vee$;
- "and" $\wedge$ and "or" $\vee$ (lattice); and
- the Boolean algebra.

[11]WHY ARE MEASUREMENTS TREATED AS PROJECTIONS? That way, we have the advantage of linearity: the projection of the vector $\vec{x}$ onto the fixed subspace $\langle \vec{e} \rangle$ with distinguished vector $\vec{e}$ is a linear function of $\vec{x}$, while the angle between the vectors is not.

[12]DUALITY. The absence of function/point duality from the undergraduate mathematics is hard to explain. It is sometimes missing even from courses in linear programming, where it has the most concrete practical meaning! From my own teaching experience (interestingly, in the physics department) I can confidently say that it is not that difficult to "dualize" the standard course of linear algebra by putting, from day one, row vectors and column vectors in two different stables: vector space V of row vectors and its dual space V^* of column vectors. Tellingly, the most convincing way to motivate the vector/covector notation is an example of a purchase of amounts g_1, g_2, g_3 of some goods at prices p^1, p^2, p^3, with the total cost being $\sum g_i p^i$. This allows us to see that the quantities g_i and p_i could be of completely different natures.

References

Mathematics: History, Philosophy, Anthropology

1. Archimedes, **The Works of Archimedes** (Sir Thomas Heath, translator). Dover Publications, 2002. (ISBN 0486420841)
2. V. I. Arnold, Polymathematics: is mathematics a single science or a set of arts? In **Mathematics: Frontiers and Perspectives** (V. I. Arnold et al., eds.). Amer. Math. Soc., 2000, pp. 403–416.
3. V. I. Arnold, **What is Mathematics? Что Такое Математика?** Издательство МЦНМО, 2004.
4. V. I. Arnold, **Yesterday and Long Ago**. Springer, 2007. ISBN 3-540-28734-5.
5. J. Azzouni, Proof and ontology in Euclidean mathematics, in **New Trends in the History and Philosophy of Mathematics** (T. H. Kjeldsen, S. A. Pedersen, and L. M. Sonne-Hanse, eds.). University Press of Southern Denmark, 2004, pp. 117–133.
6. J. Azzouni, How and why mathematics is unique as a social practice, to appear in the proceedings of PMP2002.
7. J. D. Barrow, **Pi in the Ski: Counting, Thinking and Being**. Oxford, Oxford University Press, 1987.
8. J. Barwise and J. Etchemendy, Computers, visualization, and the nature of reasoning, in **The Digital Phoenix: How Computers are Changing Philosophy** (T. W. Bynum and J. H. Moor, eds.). Blackwell, 1998, pp. 93–116.
9. D. Benson, **Music: A Mathematical Offering**, Cambridge University Press, 2006. ISBN-10: 0521619998, ISBN-13: 978-0521619998.
10. P. Bernays, Comments on Ludwig Wittgensteins Remarks on the Foundations of Mathematics, in **Philosophy of mathematics: selected readings** (P. Benacerraf and H. Putnam, eds), Prentice-Hall, Englewood Cliffs, N.J., 1964.
11. A. Bundy, M. Jamnik, and A. Fugard, What is a proof? Talk at the Royal Society Meeting "The nature of Mathematical Proof", 17 October 2004.
12. S. Clark, Tools, machines and marvels, in **Philosophy of Technology** (R. Fellows, ed.), Cambridge University Press, 1995, pp. 159–176. ISBN 0-521-55816-6.

13. M. P. Closs, ed. **Native American Mathematics**. University of Texas Press, Austin, 1988. ISBN 0-292-75537-1.

14. K. R. Coombes, AGATHOS: Algebraic geometry and thoroughly humourous overheard statements, `http://odin.mdacc.tmc.edu/~krc/agathos/aphorism.html`, accessed 12 July 2009.

15. P. J. Cohen, Comments on the foundations of set theory, Proc. Sym. Pure Math. 13, no. 1 (1971), 9-15.

16. D. Corfield, **Towards a Philosophy of Real Mathematics**. Cambridge University Press, 2003. ISBN 0521817226.

17. D. Corfield, How mathematicians may fail to be fully rational. Available at `http://www.thalesandfriends.org/en/papers/pdf/corfield_paper.pdf`, accessed 9 July 2009.

18. T. Crump, **The Anthropology of Numbers**, Cambridge University Press (Cambridge Studies in Social and Cultural Anthropology, vol. 70), 1990.

19. M. Dalbello, Mathematics for "just plain folks": The Viennese tradition of visualization of quantitative information and its verbal forms, 1889–1914. Presentation at University of Arizona, History and Philosophy of Information Access Colloquium, October 11, 2006. Available at `http://dlist.sir.arizona.edu/1585`, accessed 9 July 2009.

20. M. Dalbello and A. Spoerri, Statistical representations from popular texts for the ordinary citizen, 1889-1914. Library & Information Science Research 28, no. 1 (2006), 83-109.

21. P. Davis and R. Hersh, **The Mathematical Experience**. Birkhäuser, Boston, 1980.

22. P. Davis and R. Hersh, **Descartes' Dream**. Penguin, 1986.

23. K. Devlin, **The Math Gene: How Mathematical Thinking Evolved and Why Numbers Are Like Gossip**. Basic Books, 2000. ISBN 0-465-01618-9.

24. J. Dieudonné, **A Panorama of Pure Mathematics: As seen by N. Bourbaki**. Academic Press, New York, 1982.

25. D. C. Dennett, Memes and the exploitation of imagination. The J. Aesthetics and Art Criticism 48 (1990), 127–135.

26. D. C. Dennett, **Kinds of Minds : Towards an Understanding of Consciousness**, Phoenix Press, 2007. ISBN 0753800438.

27. K. Devlin, **Mathematics: The Science of Patterns**, Scientific American Library, 1994.

28. D. Dunham, Transformation of hyperbolic Escher patterns, `http://www.d.umn.edu/~ddunham/isis4/index.html`, accessed 12 July 2009.

29. M. Emmer, The visial mind: art, mathematics and cinema, in **The Coxeter Legacy: Reflections and Projections** (C. Davis and E. W. Ellers, eds.). American Mathematical Society, Providence, R.I., 2006, pp. 281–296.

30. P. Ernest, **The Philosophy of Mathematics Education**. Routledge Falmer, 1991. ISBN 1850006679.

31. P. Ernest, **Social Constructivism as a Philosophy of Mathematics**. State University of New York Press, 1998. ISBN 0791435873.

32. J. Fauvel, R. Flood, and R. Wilson, eds. **Music and Mathematics**, Oxford University Press, 2003.

33. C. E. Ford, The Influence of P. A. Florensky on N. N. Luzin. Historia Mathematica 25 (1998), 332-339.

34. J. Franklin, Non-deductive logic in mathematics. British J. Philosophy of Science 38 (1987), 1–18.

35. H. M. Friedman, **Philosophical Problems in Logic**. Seminar notes at the Princeton Philosophy Department, September-December, 2002, 107 pages. http://www.math.ohio-state.edu/~friedman/pdf/Princeton532.pdf. Accessed 24 October 2009.

36. M. Gardner, Is Nature ambidextrous? Philosophy and Phenomenological Research 13 (December 1952), 200–211.

37. J. Gay and M. Cole, **The New Mathematics and an Old Culture: A Study of Learning Among the Kpelle of Liberia**. Holt, Rinehart and Winston, 1967.

38. A. George and D. J. Velleman, **Philosophies of Mathematics**. Blackwell, 2002. ISBN 0–631–19544–0.

39. L. Graham and J.-M. Kantor, Russian religious mystics and French rationalists: Mathematics, 1900-1930. Bulletin of the American Academy, Spring 2005, 12–18; Isis, 97 no. 1 (March 2006) 56-74.

40. L. Graham and J.-M. Kantor, A comparison of two cultural approaches to mathematics: France and Russia, 1890-1930. Isis 97 (2006) 56-74.

41. L. Graham and J.-M. Kantor, **Naming Infinity. A True Story of Religious Mysticism and Mathematical Creativity**, Harvard University Press, 2009. ISBN 978-0-674-03293-4.

42. B. Grünbaum, What symmetry groups are present in Alhambra? Notices Amer. Math. Soc. 53, no. 6 (2006), 670–673.

43. J. Hadamard, **The Psychology of Invention in the Mathematical Field**. Dover, New York, 1945.

44. E. Hairer and G. Wanner, **Analysis by Its History**. Berlin, Springer, 1996.

45. G. H. Hardy, **A Mathematician's Apology** (with a foreword by C. P. Snow). Cambridge University Press, 1967.

46. M. Harris, "Why mathematics?" you might ask, **Princeton Companion of Mathematics** (T. Gowers, ed.), Princeton University Press, 2009, pp. 966–976.

47. T. L. Heath, **The Thirteen Books of Euclid's Elements, with Introduction and Commentary**. Cambridge, at the University Press, 1908.

48. R. Hersh, **What is Mathematics, Really?**. Vintage, London, 1998.

49. R. Hersh, Wings, not Foundations! In **Essays on the Foundations of Mathematics and Logic 1**. Advanced Studies in Mathematics and Logic, 1. Polimetrica Publisher, 2005, pp. 155–164.

50. R. Hersh, On the interdisciplinary study of mathematical practice, with a real live case study, in **Perspectives on Mathematical Practices / Bringing Together Philosophy of Mathematics, Sociology of Mathematics, and Mathematics Education** (B. van Kerkhove and J. P. van Bendegem, eds.). Springer, 2006, pp. 231–238. ISBN 140205033X.

51. W. Hodges, The geometry of music, in **Music and Mathematics** (J. Fauvel, R. Flood, and R. Wilson, eds.). Oxford University Press, 2003, pp. 91–112.

52. L. Hodgkin, **A History of Mathematics: From Mesopotamia to Modernity**. Oxford University Press, New York, 2005. ISBN 0-19-852937-6.

53. K. H. Hofmann, Commutative diagrams in the Fine Art, Notices Amer. Math. Soc. 49, no. 6 (2002), 663–668.

54. J. Hoyrup, What did the abacus teachers really do when (sometimes) ended up doing mathematics? Contribution to the conference *Perspectives on Mathematical Practices*, 26–28 March 2007, Vrije Universiteit Brussels. `http://akira.ruc.dk/~jensh/Publications/2007%7Bc%7D_What%20did%20the%20abbacus%20teachers%20really%20do.pdf`, accessed 25 October 2009.

55. J. Kepler. **Strena Seu De Nive Sexangula**, Frankfurt, 1611. English Translation: **The six-cornered snowflake** (C. Hardie, transl.). Oxford 1966.

56. D. Ker, Review of J. Fauvel, R. Flood, and R. Wilson, eds. **Music and Mathematics**, Oxford University Press, 2003. The London Math. Soc. Newsletter no. 328 (July 2004), 34–35.

57. I. Kleiner, History of the infinitely small and the infinitely large in calculus, Ed. Stud. Math. 48 (2001), 137–174.

58. D. E. Knuth, **The T$_E$Xbook**. Addison Wesley, Reading, 1984 and later editions.

59. T. Koetsier, **Lakatos' Philosophy of Mathematics: A Historical Approach**. North-Holland, Amsterdam, 1991. ISBN 0-444-88944-2.

60. S. Körner, **The Philosophy of Mathematics: An Introductory Essay**. Dover Publications, New York. ISBN 0-486-25048-2.

61. M. H. Krieger, **Doing Mathematics: Convention, Subject, Calculation, Analogy**. Singapore, World Scientific Publishing, 2003. ISBN 981-238-2062.

62. T. S. Kuhn, **The Structure of Scientific Revolutions**. Chicago, University of Chicago Press, 1962.

63. S. S. Kutateladze, Nomination and definition. `http://www.math.nsc.ru/LBRT/g2/english/ssk/nomination_e.html`. Accessed 9 July 2009.

64. I. Lakatos, **Proofs and Refutations: The Logic of Mathematical Discovery**. Cambridge University Press, 1976.

65. J. E. Littlewood, **A Mathematician's Miscellany**, Methuen & Co. Ltd., London, 1953.

66. P. J. Lu and P. J. Steinhardt, Decagonal and quasi-crystalline tilings in medieval Islamic architecture. Science, 315 (23 February 2003), 1106–1110.

67. Yu. I. Manin, **Mathematics as Metaphor.** Amer. Math. Soc., Providence, R.I., 2007. ISBN-13 978-0-8218-4331-4.

68. J.-C. Martzloff, Review of Lam Lay Yong and Ang Tian Se (1992). Fleeting footsteps: Tracing the conception of arithmetic and algebra in ancient China. Historia Mathematica 22 (1995), 67-87.

69. J. P. Miller, **Numbers in Presence and Absence: A study of Husserl's Philosophy of Mathematics.** Hague, Martinus Nijoff Publishers, 1982. ISBN 90-247-2709-X.

70. J. Mimica, **Intimations of Infinity. The Cultural Meaning of the Iqwaye Counting and Number System**. Berg Publishers, Oxford, 1992.

71. M. Mugur-Schachter and A. van der Merwe (eds.), **Quantum Mechanics, Mathematics, Cognition and Action: Proposals for a Formalized Epistemology**. Kluwer, 2003, 504 pp. ISBN-13: 978-1402011207.

72. D. Mumford, The dawning of the age of stochasticity, in **Mathematics: Frontiers and Perspectives** (V. I. Arnold et al., eds.). Amer. Math. Soc., 2000, pp. 197–218.

73. P. J. Nahin, **An Imaginary Tale. The story of** $\sqrt{-1}$. Princeton University Press, 1998.

74. Sir Isaac Newton, **Optics: Or a Treatise of the Reflections, Refractions, Inflections and Colours of Light—Based on the Fourth Edition, London, 1730** (I. Bernard Cohen et al., eds.), Dover Publications, 1952. ISBN 0486602052.

75. M. Peterson, The geometry of Piero della Francesca. Math. Intelligencer 19, no. 3 (1997), 33–40.

76. Plato, **The Collected Dialogues of Plato**. Princeton, Princeton University Press, 1961.

77. M. Raussen and C. Skau, Interview with Michael Atiyah and Isadore Singer. Notices Amer. Math. Soc. 52, no. 2 (2005), 225–233.

78. S. Restivo, **Mathematics in society and history: Sociological inquiries.** Dordrecht, Kluwer Academic Publishers, 1992.

79. E. Robson, Neither Sherlock Homes nor Babylon: a reassessment of Plimpton 322. Historia Mathematica 28 (2001), 167–206.

80. E. Robson, Words and Pictures: New Light on Plimpton 322. In **Sherlock Holmes in Babylon** (M. Anderson, V. Katz, and R. J. Wilson, eds.) Mathematical Association of America, 2004, pp. 14–26.

81. P. S. Rudman, **How Mathematics Happened. The First 50,000 Years**, Prometheus Books, Amherst, NY, 2007.

82. B. A. W. Russell, The limits of Empiricism. Proceedings of the Aristotelian Society 36 (1936), 131–150.

83. L. Russo, The definitions of fundamental geometric entities contained in Book I of Euclid's Elements. Archive for History of Exact Sciences 52, no. 3 (1998), 195–219.

84. J. A. Stedall, Of our own nation: John Wallis's account of mathematical learning in medieval England. Historia Mathematica 2 (2001), 73-122.

85. F. Viéte, **The Analytic Art** (T. R. Witmer, transl.). The Kent University Press, 1983. ISBN 0-87338-282-X.

86. U. Schwalbe and P. Walker, Zermelo and the early history of game theory. Games and Economic Behavior 34, no. 1 (2001), 123–137.

87. M. Senechal, The continuing silence of Bourbaki—An interview with Pierre Cartier, June 18, 1997. The Mathematical Intelligencer 1 (1998), 22–28.

88. S. Shapiro, **Thinking about Mathematics**. Oxford, Oxford University Press, 2000.

89. S. Shapiro, **Philosophy of Mathematics: Structure and Ontology**. Oxford University Press, 2000. ISBN 0195139305.

90. G. Sica, ed., **Essays on the Foundations of Mathematics and Logic 1.** Advanced Studies in Mathematics and Logic. Polimetrica Publisher, 2005. ISBN 88-7699-014-3.

91. J. Stillwell, **Mathematics and its History**. Springer, 1989.

92. R. Thomas, The comparison of mathematics with narrative, paper presented at "Philosophy of Mathematical Practice PMP2002", Brussels, September 2002.

93. D'A. W. Thompson, **On Growth and Form**. Cambridge University Press, 1961.

94. W. P. Thurston, On proof and progress in mathematics. Bulletin Amer. Math. Soc. 30, no. 2 (1994), 161–177.

95. F. Viéte, **The Analytic Art**, translated by T. Richard Witmer. Kent, Ohio, The Kent State University Press, 1983.

96. B. L. van der Warden, **Geometry and Algebra in Ancient Civilizations**. Springer, 1983.

97. B. L. van der Warden, **A History of Algebra**, Springer, Berlin, 1983.

98. H. Weyl, **Philosophie der Mathematik und Naturwissenschaft**. Munich, R. Oldenbourg, 1927. English translation: H. Weyl, **Philosophy of Mathematics and Natural Science**. Princeton, Princeton University Press, 1949.

99. E. Wigner, The unreasonable effectiveness of mathematics in natural sciences. Communications in Pure and Applied Mathematics 13, no. 1 (1960), 1–14.

Mathematics: Teaching

100. J. Anghileri, ed., **Children's Mathematical Thinking in the Primary Years**. Cassel, London, 1995. ISBN 0-304-33260-7.

101. P. Andrews, The curricular importance of mathematics: a comparison of English and Hungarian teachers' espoused beliefs. J. Curriculum Studies 39, no. 3 (2007), 317-338. ISSN 00220272.

102. P. Andrews, G. Hatch and J. Sayers, What do teachers of mathematics teach? An initial episodic analysis of four European traditions, in D. Hewitt and A. Noyes (eds.), **Proceedings of the Sixth British Congress of Mathematics Education Held at the University of Warwick**, 2005, pp. 9-16. `http://www.bsrlm.org.uk/IPs/ip25-1/BSRLM-IP-25-1-Full.pdf`, accessed 12 July 2009.

103. V. Arnold, *Mathematical trivium*, Soviet Math. Uspekhi 46 (1991), 225–232. (In Russian; English translations are easily available on the Internet.)

104. V. I. Arnold, On teaching mathematics, Samizdat (in Russian). Available at many websites, including `http://pauli.uni-muenster.de/~munsteg/arnold.html`, accessed 12 July 2009.

105. S. Baruk, **Échec et Maths**. Édition du Seuil, 1973. ISBN 2-02-002830-1, 2-02-004720-9.

106. H. Bass, Mathematics, mathematicians and mathematics education, Bull. Amer. Math. Soc. 42, no. 4 (2005), 417–430.

107. G. Boolos, Review of Jon Barwise and John Etchemendy, Turing's World and Tarski's World. J. Symbolic Logic 55 (1990), 370–371.

108. R. Booth and A. Borovik, Mathematics for information technology: the challenge of rigour, in **Snapshots of Innovation 2002**, Manchester, 2002, pp. 2–4. ISBN 1-903640-06-7.

109. A. V. Borovik, Implementation of the Kid Krypto concept. MSOR Connections 2, no. 3 (2002), 23–25. ISSN 1473-4869.

110. R. Brown, Promoting mathematics. MSOR Connections 7, no. 2 (2007), 24–28. ISSN 1473-4869.

111. E. Dubinsky, K. Weller, M. McDonald, and A. Brown, Some historical issues and paradoxes regarding the infinity concept: an APOS analysis, Part 1. Educational Studies in Mathematics, Educational Studies in Mathematics 58 (2005), 335–359.

112. E. Dubinsky, K. Weller, M. McDonald, and A. Brown, Some historical issues and paradoxes regarding the infinity concept: an APOS analysis, Part 2. Educational Studies in Mathematics 60 (2005), 253–266.

113. H. Freudenthal, **Mathematics as an Educational Task**. Reidel, Dordrecht, 1973.

114. A. Gardiner, The art of problem solving, in **Princeton Companion to Mathematics**, Princeton University Press, 2006.

115. D. Goldson and Steve Reeds, Using programs to teach logic to computer scientists. Notices Amer. Math. Soc. 40, no. 2 (1993), 143–148.

116. F. M. Hechinger, Learning math by thinking (from The New York Times for Tuesday, June 10, 1986), `http://www.hackensackhigh.org/math.html`.

117. R. Hersh, Independent thinking. The College Mathematics J. 34, no. 2 (2003), 112–115.

118. W. Hodges, Review of J. Barwise and J. Etchemendy, Tarski's World and Turing's World. Computerised Logic Teaching Bulletin 2, no. 1 (1989), 36–50.

119. E. Hong and Y. Aqui, Cognitive and motivational characteristics of adolescents gifted in mathematics: Comparisons among students with different types of giftedness. Gifted Child Quarterly 48, no. 3 (2004), 191–201.

120. C. Hoyles and D. Küchemann, The quality of students' explanations in geometry: insights from a large-scale longitudinal survey. `http://www.math.ntnu.edu.tw/~cyc/_private/mathedu/me1/me1_2002_1/celia.doc`, accessed 12 July 2009.

121. J. Kellermeier, Feminist pedagogy in teaching general education mathematics: creating the riskable classroom. Feminist Teacher 10, no. 1 (1996), 8–11. `http://www.tacomacc.edu/home/jkellerm/Papers/RiskableClassroom/RiskablePaper.htm`, accessed 12 July 2009.

122. V. A. Krutetskii, **The Psychology of Mathematical Abilities in Schoolchildren**. (Translated from Russian by J. Teller, edited by J. Kilpatrick and I. Wirszup.) The University of Chicago Press, 1976. ISBN 0-226-45485-1.

123. R. Lang, **Origami Design Secrets. Mathematical Methods for an Ancient Art.** A K Peters, Ltd., Natick, MA, 2003.

124. P. Liljedahl and B. Sriraman, Musings on mathematical creativity. For The Learning of Mathematics 26, no. 1 (2006), 20–23.

125. J. Mighton. **The End of Ignorance**. Alfred A. Knopf, 2007.

126. G. McColm, A metaphor for mathematical education. Notices Amer. Math. Soc. 54, no. 4 (2007), 499–502.

127. M. Montessori, **The Montessori Method** (Anne Everett George, transl.) Frederick A. Stokes Company, New York, 1912.

128. A. Ralston, Research mathematicians and mathematics education: a critique. Notices Amer. Math. Soc. 51, no. 4 (2004), 403–411.

129. B. Reznick, Some thoughts on writing for the Putnam, in **Mathematical Thinking and Problem Solving** (A. H. Schoenfeld, ed.). Lawrence Erlbaum Associates, 1994.

130. A. Sfard and P. W. Thompson, Problems of reification: representations and mathematical objects, in **Proceedings of the Annual Meeting of the International Group for the Psychology of Mathematics Education–North America, Plenary Sessions** (D. Kirshner, ed.), vol. 1. Baton Rouge LA: Louisiana State University, 1992, pp. 1–32.

131. A. Sfard, Symbolizing mathematical reality into being—or How mathematical discourse and mathematical objects create each other. In **Symbolizing and Communicating: Perspectives on Mathematical Discourse, Tools and Instructional design** (P. Cobb et al., eds.). Mahwah, NJ, Erlbaum, 1998, pp. 37-98.

132. F. Smith, **The Glass Wall: Why Mathematics May Seem Difficult**. Teachers College Press, 2002, ISBN 0-807-74241-4 (paperback), 0-807-74242-2 (cloth).

133. K. Sullivan, The teaching of elementary calculus using the nonstandard analysis approach. Amer. Math. Monthly 85, no. 5 (1976), 370-375.

134. J. E. Szydlik, Mathematical beliefs and conceptual understanding of the limit of a function. J. Res. Math. Ed. 13, no. 3 (2000), 258–276.

135. D. Tall, Reflections on APOS theory in elementary and advanced mathematical thinking, in O. Zaslavsky (ed.), **Proceedings of the 23rd Conference of PME, Haifa, Israel**. 1 (1999) pp. 111-118. http://www.warwick.ac.uk/staff/David.Tall/pdfs/dot1999c-apos-in-amt-pme.pdf, accessed 12 July 2009.

136. D. Tall, Natural and formal infinities. Educational Studies in Mathematics 48, no. 2 & 3 (2001), 199-238.

137. D. O. Tall, M. O. J. Thomas, G. Davis, E. M. Gray, and A. P. Simpson, What is the object of the encapsulation of a process? J. Mathematical Behavior 18, no. 2 (2000), 1-19.

138. P. W. Thompson, Students, functions, and the undergraduate mathematics curriculum, in **Research in Collegiate Mathematics Education** (E. Dubinsky et al., eds.). (Issues in Mathematics education, vol. 4). American Mathematical Society, Providence, RI, 1994, pp. 21–44.

139. W. P. Thurston, Mathematical education. Notices Amer. Math. Soc. 37 (1990), 844–850.

140. T. Tinsley, The use of origami in the mathematics education of visually impaired students. Education of the Visually Handicapped 4, no. 1 (1972), 8–11.

141. J. Vlassis, Making sense of the minus sign or becoming flexible in 'negativity'. Learning and Instruction 14 (2004), 469-484.

142. K. Weller, A. Brown, E. Dubinsky, M. McDonald, and C. Stenger, Intimations of infinity. Notices Amer. Math. Soc. 51, no. 7 (2004), 741–750.

143. D. Wright, **Mathematics and Music**. Amer. Math. Soc., Providence, RI, 2009. ISBN-13: 978-0821848739.

144. A. Zvonkin, Mathematics for little ones. J. Mathematical Behavior 11, no. 2 (1992), 207–219.

145. A. Zvonkin, Children and $\binom{5}{2}$. J. Mathematical Behavior 12, no. 2 (1993), 141–152.

Cognitive Science, Psychology, Neurophysiology

146. K. Akins, What is it to be boring and myopic? In **Dennett and His Critics** (B. Dalhobon, ed.). Blackwell Publishers, Cambridge, MA, 1993, pp. 124–160.

147. O. Akman, D. S. Broomhead, and R. A. Clement, Mathematical models of eye movements. Mathematics Today 39, no. 2 (2003), 54–59.

148. M. W. Alibali and A. A. DiRusso, The function of gesture in learning to count: More than keeping track. Cognitive Development 14 (1999), 37-56.

149. A. J. Anderson and P. W. McOwan, Model of a predatory stealth behaviour camouflaging motion. Proc. Roy. Soc. London B 270 (2003), 189-195.

150. A. J. Anderson and P. W. McOwan, Humans deceived by predatory stealth strategy camouflaging motion. Proc. Roy. Soc. London B (Suppl.) Biology Letters 03b10042.S1–03b10042.S3.

151. S. E. Antell and D. P. Keating, Perception of numerical invariance in neonates. Child Development 54 (1983), 697-701.

152. M. H. Ashcraft, and J. A. Krause, Working memory, math performance, and math anxiety. Psychonomic Bulletin Review 14, no. 2 (2007), 243–248.

153. D. H. Ballard, M. M. Hayhoe, P. K. Pook, and R. P. N. Rao, Deictic codes for the embodiment of cognition, http://www.bbsonline.org/Preprints/OldArchive/bbs.ballard.html, accessed 12 July 2009.

154. S. Baron-Cohen, **The Essential Difference. The Truth about the Male and Female Brain**. Basic Books, New York, 2003.

155. S. Blackmore, **The Meme Machine**. Oxford University Press, 1999.

156. P. Blenkhorn and D. G. Evans, Using speech and touch to enable blind people to access schematic diagrams. J. Network and Computer Applications 21, no. 1 (1998), 17–29.

157. D. S. Broomhead, R. A. Clement, M. R. Muldoon, J. P. Whittle, C. Scallan, and R. V. Abadi, Modelling of congenital nystagmus waveforms produced by saccadic system abnormalities. Biological Cybernetics 82, no. 5 (May 2000), 391–399.

158. P. Bryant and S. Squire, The influence of sharing on children's initial concept of division. J. Experimental Child Psychology 81, no. 1 (January 2002), 1–43.

159. B. Butterworth, **The Mathematical Brain**. Papermac, London, 1999.

160. B. Butterworth, Mathematics and the brain: Opening address to the Mathematics Association, Reading, UK, April 3rd 2003, www.mathematicalbrain.com/pdf/.

161. S. Carey, Cognitive foundations of arithmetic: foundations and ontogenesis. Mind and Language 16, no. 1 (February 2001), 37–55.

162. S. Carey, Bootstrapping and the origin of concepts. Dedalus, winter 2004, 59–68.

163. S. Carey and B. W. Sarnecka, The development of human conceptual representations: A case study, in **Processes of Change in Brain and Cognitive Development. Attention and Performance XXI**

(Y. Munakata and M. H. Johnson, eds.). Oxford University Press, 2004, pp. 473-496.

164. J.-P. Changeux, **Neuronal Man: The Biology of Mind**. Princeton University Press, 1997. ISBN 0-691-02666-1.

165. H. G. Le Clec, S. Dehaene, C. Cohen, J. Mehler, E. Dupoux, J. B. Poline, S. Lehericy, P. F. van de Moortele, and D. Le Bihan, Distinct cortical areas for names of numbers and body parts independent of language and input modality. NeuroImage 12 (2000), 381-391.

166. L. A. Cooper and R. N. Shepard, Chronometric studies of the rotation of mental images, in **Visual Information Processing** (W. G. Chase, ed.). New York, Academic Press, 1973, pp. 75-176.

167. R. Dawkins, **The Selfish Gene**. Oxford, Oxford University Press, 1976 (new edition with additional material, 1989).

168. R. Dawkins, *Foreword to* S. Blackmore, **The Meme Machine**. Oxford University Press, 1999, pp. vii-xvii.

169. H. De Cruz, Why are some numerical concepts more successful than others? An evolutionary perspective on the history of number concepts. Evolution and Human Behavior 27 (2006), 306-323.

170. S. Dehaene, E. Spelke, P. Pinet, R. Stanescu and S. Tsivkin, Sources of mathematical thinking: behavioral and brain–imaging evidence. Science 284 (7 May 1999), 970–974.

171. S. Dehaene, **The Number Sense**. Penguin Books, 2001.

172. S. Dehaene, V. Izard, P. Pica, and E. S. Spelke, Core knowledge of geometry in an Amazonian indigene group. Science 311 (2005), 381–384.

173. T. Diggle, S. Kurniawan, D. G. Evans, and P. Blenkhorn, An Analysis of layout errors in word processed documents produced by blind people, in **Computer Helping People with Special Needs : 8th International Conference, ICCHP 2002, Linz, Austria, July 15-20, 2002**. Lect. Notes Comp. Sci. 2398 (2002), 587–588.

174. K. Distin, **The Selfish Meme**, Cambridge University Press, 2005. ISBN 0-521-60627-6.

175. G. M. Edelman, **Neural Darwinism: The Theory of Neuronal Group Selection**. Basic Books, New York, 1987.

176. E. Eger, P. Sterzer, M. O. Russ, A.-L. Giraud, and A. Kleinschmidt, A supramodal number representation in human intraparietal cortex. Neuron 37 (2003), 1-20.

177. D. L. Everett, Cultural constraints on grammar and cognition in Pirahã: Another look at the design features of human language. Current Anthropology 46, no. 4 (2005), 621–646.

178. G. J. Feist, **The Psychology of Science and the Origins of the Scientific Mind**. Yale University Press, 2006, xx + 316 pp.

179. M. H. Fischer, Cognitive representation of negative numbers. Psychological Science 14 (2003), 278-282.

180. C. R. Gallistel and R. Gelman, Non-verbal numerical cognition: From reals to integers. Trends in Cognitive Sciences 4 (2000), 59-65.

181. D. C. Geary, Reflections of evolution and culture in children's cognition. Implications for mathematical development and instruction. American Psychologist 50 (1995), 24-37.

182. I. J. Gilchrist, V. Brown, and J. Findlay, Saccades without eye movements. Nature 390, no. 3 (1997), 130–131.

183. P. Glendinning, View from the Pennines: Non-trivial pursuits. Mathematics Today 39, no. 4 (August 2003), 118–120.

184. P. Glendinning, The mathematics of motion camouflage. Proc. Roy. Soc. (London) Series B 271 (2004), 477–481.

185. P. Glendinning, View from the Pennines: seven plus or minus two. Mathematics Today 41, no. 6 (December 2005), 189–190.

186. P. Gordon, Numerical cognition without words: Evidence from Amazonia. Science 306 (2004), 496-499.

187. B. Greer, The growth of mathematics through conceptual restructuring. Learning and Instruction 14 (2004), 541-548.

188. R. Gregory, **Eye and Brain: The Psychology of Seeing**, 5th ed., Princeton University Press, 1998.

189. M. D. Hauser, S. Carey, and L. B. Hauser, Spontaneous number representation in semi-free-ranging rhesus monkeys. Proceedings of the Royal Society London B 267 (2000), 829-833.

190. W. Heiligenberg, **Principles of Electrolocation and Jamming Avoidance Studies of Brain Function**. Vol 1. Berlin: Springer-Verlag, 1977.

191. D. W. Henderson, Review of *Where Does Mathematics Come From* by Lakoff and Nunez. Mathematical Intelligencer 24, no. 1 (2002), 75–78.

192. B. Hermelin and N. O'Connor, Idiot savant calendrical calculators: rules and regularities. Psychol Med. 16, no. 4 (1986), 885–893.

193. L. Hermer and E. S. Spelke, A geometric process for spatial reorientation in young children. Nature 370, (1994) 57–59.

194. D. Hofstadter, **I am a Strange Loop**. Basic Books, New York, 2007. ISBN-13: 978-0-465-03078-1.

195. N. Humphrey, Caveart, autism, and the evolution of human mind. Cambridge Archeological J. 8, no. 2 (1998), 165–191.

196. J. Hurford, Languages Treat 1–4 Specially. Mind and Language 16 (2001), 69.

197. S. Ishihara and K. Kaneko, Magic number 7 ± 2 in network of treshhold dynamics. Physical Review Letters 94 (2005), 058102.

198. K. Kaneko, Dominance of Milnor attractors in globally coupled dynamical systems with more than $7\pm$ degrees of freedom. Physical Review E. 66 (2002), 055201(R).

199. D. Kimura, Sex Differences in the Brain. Scientific American May 13, 2002. http://scientificamerican.com/article.cfm?articleID=00018E9D-879D-1D06-8E49809EC588EEDF.

200. G. Lakoff and R. Núnez, **Where Mathematics Comes From: How the Embodied Mind Brings Mathematics Into Being**. Basic Books, New York, 2000.

201. V. A. Lefebvre. **Algebra of Conscience**. Revised Edition with a new Foreword by Anatol Rapoport. Series: Theory and Decision Library A: Vol. 30 (2001), 372 pp. ISBN: 0-7923-7121-6.

202. A. N. Leontiev, **Проблемы развития психики**, Издательство МГУ, Москва, 1981 (in Rusian).

203. M. Lotze, G. Scheler, H.-R. M. Tan, C. Braun, and N. Birbaumer, The musician's brain: functional imaging of amateurs and professionals during performance and imagery. NeuroImage 20 (2003), 1817–1829.

204. A. R. Luria, **Потерянный и возвращенный мир.** Издательство МГУ, Москва, 1971 (in Rusian). The following seems to be an English translation of the book: **The Man with a Shattered world: The History of a Brain Wound**. Harvard University Press, Cambridge, MA, 1987.

205. K. Menninger, **Number Words and Number Symbols: A Cultural History of Numbers**. MIT Press, 1970.

206. G. A. Miller, The magical number seven, plus or minus two: Some limits on our capacity for processing information. The Psychological Revue 63 (1956), 81–97.

207. A. Mizutani, J. S. Chahl, and M. V. Srinivasen. Motion camouflage in dragonflies, Nature 423 (2003), 604.

208. R. S. Moyers and T. K. Landauer, Time required for judgements of numerical inequality. Nature 215 (1967), 1519-1520.

209. M. E. Nelson, Target detection, image analysis and modeling, in **Electroreception** (Springer Handbook of Auditory Research) (T. H. Bullock, C. D. Hopkins, A. N Popper, and R. R. Fay, eds.). Springer, New York, 2005, pp. 290–317.

210. A. Nieder and E. K. Miller, Coding of cognitive magnitude: Compressed scaling of numerical information in the primate prefrontal cortex. Neuron 37 (2003), 149-157.

211. T. Nørretranders, **The User Illusion: Cutting Consciousness Down to Size**. Penguin 1998. ISBN 0-713-99182-8.

212. D. Norris, How to build a connectionist idiot (savant). Cognition 35, no. 3 (1990), 277–291.

213. R. Núñez, Do *real* numbers really move? Language, thought and gesture: the embodied cognitive foundations of mathematics, in **18 Unconventional Essays on the Nature of Mathematics** (R. Hersh, ed.). Springer, 2005, pp. 160–181.

214. N. O'Connor and B. Hermelin, Idiot savant calendrical calculators: maths or memory? Psychol Med. 14, no. 4 (1984), 801–806.

215. S. Peterson and T. J. Simon, Computational evidence for the subitizing phenomenon as an emergent property of the human cognitive architecture, Cognitive Science 24, no. 1 (2000), 93–122.

216. M. Piazza, A. Mechelli, B. Butterworth, and C. J. Price, Are subitizing and counting implemented as separate or functionally overlapping processes? NeuroImage 15 (2002), 435–446.

217. P. Pica, C. Lemer, V. Izard, and S. Dehaene, Exact and approximate arithmetic in an Amazonian indigene group. Science 306 (2004), 499-503.

218. S. Pinker, **The Language Instinct**. Penguin Books, 1995.

219. S. Pinker, **How the Mind Works**. Penguin Books, 1999.

220. V. Ramachandran, **The Emerging Mind**. Profile Books, 2003. ISBN 1-86197-303-9.

221. B. Rasnow and J. M. Bower, Imaging with electricity: how weakly electric fish might perceive objects, in **Computational Neuroscience Trends in Research 1997**. Plenum, New York, 1997, pp. 795–800.

222. A. F. da Rocha and E. Massad, How the human brain is endowed for mathematical reasoning. Mathematics Today 39, no. 3 (2003), 81–84.

223. O. Sacks, **The Man who Mistook his Wife for a Hat**. Duckworth, London, 1985.

224. B. W. Sarnecka and S. A. Gelman, Six does not just mean a lot: Preschoolers see number words as specific. Cognition 92 (2004), 329-352.

225. B. W. Sarnecka, V. G. Kamenskaya, Y. Yamana, T. Ogura, and J. B. Yudovina, From grammatical number to exact numbers: Early meanings of one, 'two', and 'three' in English, Russian, and Japanese. Cognitive Psychology (2006), doi:10.1016/j.cogpsych.2006.09.001.

226. G. B. Saxe, Body parts as numerals: A developmental analysis of numeration among the Oksapmin in Papua New Guinea. Child Development 52 (1981), 306-316.

227. R. Sekuler and R. Blake, **Perception**. 4th edition, McGraw-Hill, 2002.

228. L. Selfe, **Nadia: a Case of Extraordinary Drawing Ability in Children.** Academic Press, London, 1977. ISBN 9780126357509.

229. N. J. Smelser and P. B. Baltes, **International Encyclopedia of the Social and Behavioral Sciences**. Amsterdam, New York: Elsevier, 2001. ISBN 0-08-043076-7.

230. O. Simon, J.-J. Mangin, L. Cohen, D. Le Bihan, and S. Dehaene, Topographical layout of hand, eye, calculation, and language-related areas in the human parietal lobe. Neuron 33 (2002), 475-487.

231. A. W. Snyder and D. J. Mitchell, Is integer arithmetic fundamental to mental processing?: The mind's secret arithmetic. Proc. Roy. Soc. London B 266 (1999), 587–592.

232. M. V. Srinivasan and M. Davey, Strategies for active camouflage of motion. Proc. Roy. Soc. London B 259 (1995), 19-25.

233. M. J. Tarr, Rotating objects to recognise them: A case study on the role of viewpoint dependency in the recognition of three dimensional shapes. Psychonomic Bulletin and Review 2 (1995), 55–82.

234. M. J. Tarr and M. J. Black, A computational and evolutionary perspective on the role of representation in vision. Computer Vision, Graphics, and Image Processing: Image Understanding 60 (1994), 65–73.

235. M.J. Tarr and S. Pinker, Mental rotation and orientation-dependence in shape recognition. Cognitive Psychology 21 (1989), 233–282.

236. E. Temple and M. I. Posner, Brain mechanisms of quantity are similar in 5-year-old children and adults. Proc. Nat. Acad. Sci. USA 95 (1998), 7836-7841.

237. C. Uller, M. Hauser, and S. Carey, Spontaneous representation of number in cotton-top tamarins (*Saguinus oedipus*). J. Comparative Psychology 115 (2001), 248-257.

238. C. Uller, R. Jaeger, G. Guidry, and C. Martin, Salamanders (*Plethodon cinereus*) go for more: Rudiments of number in an amphibian. Animal Cognition 6 (2003), 105-112.

239. L. R. Vandervert, A measurable and testable brain-based emergent interactionism: An alternative to Sperry's mentalist emergent interactionism. J. Mind and Behavior 12, no. 2 (1991), 210–220. ISSN 0271-0137.

240. L. R. Vandervert, Neurological positivism's evolution of mathematics. J. Mind and Behavior 14, no. 3 (1993), 277–288. ISSN 0271-0137.

241. L. R. Vandervert, How the brain gives rise to mathematics in on-togeny and in culture. J. Mind and Behavior 15, no. 4 (1994), 343–350. ISSN 0271-0137.

242. L. R. Vandervert, From idiots savants to Albert Einstein: A brain-algorithmic explanation of savant and everyday performance. New Ideas in Psychology 14, no. 1 (March 1996), 81–92.

243. L. R. Vandervert, A motor theory of how consciousness within language evolution led to mathematical cognition: origin of mathematics in the brain. New Ideas in Psychology 17, no. 3 (1999), 215–235.

244. R. A. Varley, N. J. C. Klessinger, C. A. J. Romanowski, and M. Siegal, Agrammatic but numerate. Proc. Nat. Acad. Sci. USA 102 (2005), 3519-3524.

245. H. M. Wellman and K. F. Miller, Thinking about nothing: Development of concepts of zero. British J. Developmental Psychology 4 (1986), 31-42.

246. L. A. White, The locus of mathematical reality: an anthropological footnote, in **18 Unconventional Essays on the Nature of Mathematics** (R. Hersh, ed.). Springer, 2005, pp. 304–319.

247. K. Wynn, Addition and subtraction by human infants. Nature 358 (1992), 749-750.

248. K. Wynn, Psychological foundations of number: Numerical competence in human infants. Trends in Cognitive Sciences 2 (1998), 296-303.

249. K. Wynn and W. Chiang, Limits to infants knowledge of objects: The case of magical appearance. Psychological Science 9 (1998), 448-455.

250. F. Xu and E. S. Spelke, Large number discrimination in 6-month-old infants. Cognition 74 (2000), B1-B11.

251. H. Zenil and F. Hernandez-Quiroz, On the possible computational power of the human mind, arXiv:cs.NE/0605065, v3, 11 Jun 2006.

Popular Mathematics and Popular Computer Science

252. R. C. Alperin, Heron's Area Formula. The College Mathematics J. 18, no. 2 (1987), 137–138.

253. R. C. Alperin, A mathematical theory of origami constructions and numbers. New York J. Math. 6 (2000), 119–133, http://nyjm.albany.edu:8000/j/2000/6-8.html.

254. R. C. Alperin, Mathematical origami: another view of Alhazen's optical problem, in **Origami³: Third International Meeting of Origami Science, Mathematics, and Education** (T. Hull, ed.). A K Peters, Natick, Massachusetts, 2002, pp. 83–93.

255. R. C. Alperin, Trisections and totally real origami, www.math.sjsu.edu/~alperin/TRFin.pdf, accessed 12 July 2009.

256. A. Ash and R. Gross, **Fearless Symmetry: Exposing the Hidden Patterns of Numbers**. Princeton University Press, Princeton, 2006.

257. R. Ball, The Kolmogorov cascade, an entry in **Encyclopedia of Nonlinear Science** (A. Scott, ed.). Routledge, 2004. The text is available at http://wwwrsphysse.anu.edu.au/~rxb105/cascade.pdf, accessed 12 July 2009.

258. M. P. Beloch, Sulla risoluzione dei problemi di terzo e quarto grado col metodo del ripiegamento della carta, Scritti Matematici Offerti a Luigi Berzolari, Pavia, 1936, pp. 93-96.

259. E. Berlekamp, J. H. Conway, and R. Guy, **Winning Ways for Your Mathematical Plays**. Academic Press, New York, 1982.

260. W. Casselman, If Euclid had been Japanese. Notices Amer. Math. Soc. 54, no. 5 (2007), 626–628.

261. E. Castronova, **Synthetic Worlds**. University of Chicago Press, 2005. ISBN 0-226-09626-2.

262. R. J. McG. Dawson, On removing a ball without disturbing the others. Math. Mag. 57 (1984), 27–30.

263. R. W. Doerfler, **Dead Reckoning. Calculating without Instruments.** Gulf Publishing Company, Houston, 1993. ISBN 0-88415-087-9.

264. H. Dörrie, **100 Great Problems of Elementary Mathematics**. Dover Publications, New York, 1965.

265. E. B. Dynkin, S. A. Molchanov, A. P. Rozental, and A. N. Tolpygo, **Mathematical Problems**. Moscow, Nauka, 1971.

266. I. M. Gelfand, E. G. Glagoleva, and A. A. Kirillov, **The Method of Coordinates**. Birkhäuser, Boston, 1990.

267. I. M. Gelfand, E. G. Glagoleva, and E. Shnol, **Functions and Graphs**. Birkhäuser, Boston, 1990.

268. V. Gutenmacher and N. B. Vasilyev, **Lines and Curves**, Birkhäuser, Boston, 2004.

269. L. Habsieger, M. Kazarian, and S. Lando, On the second number of Plutarch. Amer. Math. Monthly 105 (1998), 446.

270. L. D. Henderson, **The Fourth Dimension and Non-Euclidean Geometry in Modern Art**. Princeton University Press, 1983.

271. D. W. Henderson, **Experiencing Geometry in Euclidean, Spherical, and Hyperbolic Spaces**. Prentice Hall, 2001.

272. D. W. Henderson and D. Taimina, **Experiencing Geometry: Euclidean and Non-Euclidean With History**, 3rd ed., Prentice-Hall, 2004.

273. D. W. Henderson and D. Taimina, Crocheting the Hyperbolic Plane. Mathematical Intelligencer 23, no. 2 (2001), 17–28.

274. R. Hersh, A nifty derivation of Heron's Area Formula by 11th Grade Algebra. Focus 22, no. 8 (2002), 22–22.

275. C. H. Hinton, **The Fourth Dimension**, George Allen & Co., London, 1904 and later editions.

276. C. Hobbs and R. Perryman, **The Largest Number Smaller Than Five**. Lulu, 2007, 123 pp. ISBN 978-1-4303-0630-6.

277. T. Hull, **Project Origami : Activities for Exploring Mathematics**. A K Peters, 2006. ISBN 1-56881-258-2.

278. H. Huzita, A problem on the Kawasaki theorem, in **Proceedings of the First International Meeting of Origami Science and Technology** (H. Huzita, ed.). 1989, pp. 159–163.

279. H. Huzita, Axiomatic development of origami geometry, **Proceedings of the First International Meeting of Origami Science and Technology** (H. Huzita, ed.). 1989, pp. 143–158.

280. H. Huzita, The trisection of a given angle solved by the geometry of origami, in **Proceedings of the First International Meeting of**

Origami Science and Technology, (H. Huzita, ed.). 1989, pp. 195–214.

281. H. Huzita, Understanding Geometry through Origami Axioms, in **Proceedings of the First International Conference on Origami in Education and Therapy (COET91)** (J. Smith, ed.). British Origami Society, 1992, pp. 37–70).

282. H. Huzita, Understanding geometry through origami axioms: is it the most adequate method for blind children?, in **Proceedings of the First International Conference on Origami in Education and Therapy (COET91)** (J. Smith, ed.). British Origami Society, 1992, pp. 37–70.

283. H. Huzita, Drawing the regular heptagon and the regular nonagon by origami (paper folding). Symmetry: Culture and Science 5, no. 1 (1994), 69–84.

284. H. Huzita and B. Scimemi, The algebra of paper-folding (origami), **Proceedings of the First International Meeting of Origami Science and Technology** (H. Huzita, ed.). 1989, pp. 215-222.

285. D. A. Klain, An intuitive derivation of Heron's formula, `http://faculty.uml.edu/dklain/klain-heron.pdf`, accessed 12 July 2009.

286. H. Osinga and B. Krauskopf, Crocheting the Lorenz manifold. Mathematical Intelligencer 26, no. 4 (2004), 25–37.

287. B. Scimemi, Algebra og geometri ved hjep av papirbretting (Algebra and geometry by folding paper). Translated from the Italian with comments by Christoph Kirfel. Nordisk Maematisk Tidsskrift 4 (1988).

288. E. S. Selmer, Registration numbers in Norway: some applied number theory and psychology. J. Roy. Statist. Soc. 130 (1967), 225–231.

289. R. P. Stanley, Hipparchus, Plutarch, Schröder and Hough. Amer. Math. Monthly 104 (1997), 344–350.

290. I. Stewart, **Does God play dice? The mathematics of chaos**. Penguin, London, 1990.

291. A. S. Tarasov, Solution of Arnol′d's "folded ruble" problem. (Russian) ChebyshevskiĭSb. 5, no. 1(9) (2004), 174–187.

292. V. Ufnarovski (В. Уфнаровский), **Математический Аквариум**. Научно-издательский центр "Регулярная и хаотическая динамика", Moscow, 2000, ISBN 5-89806-030-8.

Mathematics Textbooks

293. J. Barwise and J. Etchemendy, **The Language of First-Order Logic: Including the IBM-compatible Windows Version of Tarski's World 4.0**. Stanford, 1992.

294. A. V. Borovik and A. Borovik, **Mirrors and Reflections**. Birkhäuser, Boston, 2009.

295. G. Choquet, **L'enseignement de la géométrie**. Hermann, 1964. (English translation: **Geometry in a Modern Setting**. Hermann, 1969.)

296. H. S. M. Coxeter, **Introduction to Geometry**. John Wiley & Sons, New York, London, 1961 and later eds.

297. J. Hadamard, **Leçons de géométrie élementaire**. BiblioBazaar, 2009. ISBN-13 978-1110978281.

298. R. Hartshorne, **Foundations of Projective Geoemetry**. New York, W. A. Benjamin, 1967.

299. R. Hartshorne, **Geometry: Euclid and Beyond**. Springer, 2000.

300. I. N. Herstein, **Topics in Algebra**. Braisdell Publishing Company, New York, 1964.

301. R. Hill, **A First Course in Coding Theory**. Clarendon Press, 1986. ISBN 0-19-853803-0.

302. A. Holme, **Geometry: Our Cultural Heritage**. Springer, 2002.

303. U. H. Kortenkamp and J. Richter-Gebert. **The Interactive Geometry Software Cinderella**. Springer-Verlag, 1999. ISBN 3540147195.

304. D. Poole, **Linear Algebra: A Modern Introduction**. Thompson, 2006.

305. F. D. Porturaro and R. E. Tully, **Logic with Symlog: Learning Symbolic Logic by Computer**. Prentice Hall, 1994.

306. D. M. Y. Sommerville, **Analytical Geometry of Three Dimensions**. Cambridge University Press, 1951.

Hardcore Mathematics, Computer Science, Mathematical Biology

307. V. I. Arnold, О классах когомологий алгебраических функций, сохраняющихся при преобразованиях Чирнгаузена. Functional Analysis and its Application. 4, no. 1 (1970), 84–85 (in Russian); English translation: The cohomology classes of algebraic functions invariant under Tschirnhausen transformations, Functional Anal. Appl. 4 (1970), 74–75.

308. V. I. Arnold, First steps of local contact algebra, Canad. J. Math. 51 (6) (1999), 1123-1134.

309. F. Bachmann, **Aufbau der Geometrie aus dem Spiegelungsbegriff**, Springer-Verlag, Berlin-Göttingen-Heidelberg, 1959.

310. R. Baire, **Leçons sur les Fonctions Discontinues**. Paris, 1904.

311. D. J. Benson, **Polynomial Invariants of Finite Groups**. (London Math. Soc. Lect. Notes Ser. 190.) Cambridge University Press, 1993, 128 pp. ISBN: 0521458862.

312. I. F. Blake, G. Seroussi, and N. Smart, **Elliptic Curves in Cryptography**. London Mathematical Society Lecture Note Series. Cambridge University Press, 1999. ISBN 0521653746.

313. A. Blass, Yu. Gurevich, and S. Shelah, Choiceless polynomial time. Annals of Pure and Applied Logic 100 (1999), 141–187.

314. A. Blass, Yu. Gurevich, and S. Shelah, On polynomial time computation over unordered structures. J. Symbolic Logic 67, no. 3 (2002), 1093–1125.

315. A. Blass and Yu. Gurevich, Strong extension axioms and Shelah's zero-one law for choiceless polynomial time. J. Symbolic Logic 68, no. 1 (2003), 65–131.

316. A. Blinder, Wage discrimination: reduced form and structural estimates. J. Human Resources 8 (1973), 436–455.

317. R. F. Booth, D. Y. Bormotov, and A. V. Borovik, Genetic algorithms and equations in free groups and semigroups, in **Computational and Experimental Group Theory**. Contemp. Math. 349 (2004), 63–81.

318. R. F. Booth and A. V. Borovik, Coevolution of algorithms and deterministic solutions of equations in free groups, in **Genetic Programming. 7th European Conference, EuroGP 2004. Coimbra, Portugal, April 2004** (M. Keijzer et al., eds.). Lecture Notes Comp. Sci., vol. 3003, Springer-Verlag, 2004, pp. 11–22.

319. Z. I. Borevich and I. R. Shafarevich, **Number Theory**. З. И. Боревич и И. Р. Шафаревич, **Теория Чисел**. Москва, Наука, 1972 (in Russian).

320. A. V. Borovik, I. M. Gelfand, and N. White, **Coxeter Matroids**. Birkhäuser, Boston, 2003.

321. A. V. Borovik and A. Nesin, **Groups of Finite Morley Rank**. Oxford University Press, Oxford, 1994.

322. J. M. Borwein, Multi-variable sinc integrals and the volumes of polyhedra. Algorithms Seminar 2001-2002, F. Chyzak (ed.). INRIA, 2003, pp. 8992. Available online at http://algo.inria.fr/seminars/sem01-02/borwein1.pdf, accessed 9 July 2009.

323. N. Bourbaki, **Groupes et Algebras de Lie, Chap. 4, 5, et 6**. Hermann, Paris, 1968.

324. J. Buchmann and H. C. Williams, A key-exchange system based on imaginary quadratic fields. J. Crypto. 1 (1988), 107–118.

325. A. Bundy, D. Basin, D. Hutter, and A. Ireland, **Rippling: Meta-Level Guidance for Mathematical Reasoning**. Cambridge University Press, 2005.

326. P. Cameron, The random graph revisited, in **European Congress of Mathematics, Barcelona, July 10-14, 2000**, Volume II (C. Casacuberta, R. M. Miró-Roig, J. Verdera, and S. Xambó-Descamps eds.). Birkhäuser, Basel, 2001, pp. 267–274.

327. B. Casselman, **Mathematical Illustrations: A Manual of Geometry and Postscript**. Cambridge University Press, 2004. ISBN 0-521-54788-1.

328. G. Chaitin, The halting probability Omega: Irreducible complexity in pure mathematics. Milan J. Mathematics, 75 (2007); http://www.cs.umaine.edu/~chaitin/mjm.html, accessed 12 July 2009.

329. S. Chari, C. S. Jutla, J. R. Rao, and P. Rohatgi, Towards sound approaches to counteract power-analysis attacks, in **Advances in Cryptography: Proc. of Crypto '99**. Springer-Verlag, Lect. Notes Comp. Sci. 1666 (1999), 398–412.

330. G. L. Cherlin and B. J. Latka, Minimal antichains in well-founded quasiorders, and tournaments. J. Combinatorial Theory Ser. B 80 (2000), 258–276

331. A. J. Chorin, Book Review: Kolmogorov spectra of turbulence I: Wave turbulence, by V. E. Zakharov, V. S. Lvov, and G. Falkovich. Bull. Amer. Math. Soc. 29, no. 2 (1993), 304-306.

332. J. Copeland, Even Turing machines can compute uncomputable functions, in **Unconventional Models of Computation (Discrete Mathematics and Theoretical Computer Science)** (C. Calude, J. Casti and M. J. Dinneen, Eds.). London, Springer-Verlag, 1998, pp. 150–164.

333. J. Copeland, Accelerating Turing machines. Minds and Machines 12 (2002), 281-301.

334. J. Copeland, Hypercomputation in the Chinese Room, in **Unconventional Models of Computation 2002** (C.S. Calude et al. eds.). Lect. Notes Comp. Sci. 2509 (2002), 15-26.

335. J. P. Costas, A study of a class of detection waveforms having nearly ideal range-Doppler ambiguity properties. Proceedings of the IEEE, 72 no. 9 (1984), 996–1009.

336. H. S. M. Coxeter, Finite groups generated by reflections and their subgroups generated by reflections. Proc. Camb. Phil. Soc. 30 (1934), 466–482.

337. H. S. M. Coxeter, The complete enumeration of finite groups of the form $R_i^2 = (R_i R_j)^{k_{ij}} = 1$. J. London Math. Soc. 10 (1935), 21–25.

338. H. S. M. Coxeter, **Regular Polytopes**. Methuen and Co., London, 1948.

339. H. T. Croft, K. J. Falconer, and R. K. Guy, **Unsolved Problems in Geometry**. Springer-Verlag, New York, 1991.

340. E. D. Demaine and J. B. Mitchell, Reaching folded states of a rectangular piece of paper. In Proc. 13th Canadian Conf. Comput. Geom., pp. 7375, 2001.

341. E. D. Demaine, S. L. Devadoss, J. S. B. Mitchell, and J. O'Rourke, Continuous Foldability of Polygonal Paper. Proceedings of the 16th Canadian Conference on Computational Geometry.

342. E. D. Demaine and J. O'Rourke, **Geometric folding algorithms: linkages, origami, polyhedra**. Cambridge University Press, 2007. ISBN 9780521857574.

343. J. Denef and F. Loeser, Germs of arcs on singular algebraic varieties and motivic integration. Invent. Math. 135, no. 1 (1999), 201–232.

344. J. Denef and F. Loeser, Definable sets, motives and p-adic integrals. J. Amer. Math. Soc. 14, no. 2 (2000), 429–469.

345. P. Doyle and J. Conway, Division by three, http://arxiv.org/abs/math.LO/0605779, accessed 12 July 2009.

346. A. V. Dyskin, Y. Estrin, A. J. Kanel-Belov, and E. Pasternak, A new concept in design of materials and structures: Assemblies of interlocked tetrahedron-shaped elements. Scripta Materialia 44 (2001), 2689–2694.

347. A. V. Dyskin, Y. Estrin, A. J. Kanel-Belov, and E. Pasternak, Toughening by fragmentation – How topology helps. Advanced Engineering Materials, 3, Issue 11 (2001), 885–888.

348. A. V. Dyskin, Y. Estrin, A. J. Kanel-Belov, and E. Pasternak, Topological interlocking of platonic solids: A way to new materials and structures. Phil. Mag. Letters 83, no. 3 (2003), 197–203.

349. D. B. A. Epstein, with J. W. Cannon, D. F. Holt, S. V. F. Levy, M. S. Paterson, and W. P. Thurston, **Word Processing in Groups**. Jones and Bartlett, Boston-London, 1992.

350. L. Fejes Toth and A. Heppes, Uber stabile Korpersysteme, Composito Math. 15 (1963), 119–126.

351. S. Fomin and A. Zelevinsky, Y-systems and generalized associahedra, Adv. Math. 158 (2003) 977–1088.

352. S. Forrest, Openning talk at EuroGP2004, 5 April 2004, Coimbra, Portugal.

353. S. Forrest, J. Balthrop, M. Glickman, and D. Ackley, Computation in the wild, in **The Internet as a Large-Scale Complex System** (K.

Park and W. Willins, eds.). Oxford University Press, 2005, 322 pp. ISBN 0195157206.

354. H. Gelernter, A note on syntactic symmetry and the manipulation of formal systems by machine. Information and Control 2 (1959), 80–89.

355. M. Glickman, The G-block system of vertically interlocked paving, in **Proc. Second Internat. Conf. on Concrete Block Paving**. Delft, April 10–12, 1984, pp. 345–348.

356. S. W. Golomb, Algebraic constructions for Costas arrays. J. Combin. Theory Ser. A 37 (1984), 13–21.

357. S. W. Golomb, The status of Costs array constructions, in **Proceedings of the 40th Annual Conference on Information Sciences and Systems**. Princeton, NJ, March 2006.

358. S. W. Golomb and H. Taylor, Construction and properties of Costas arrays. Proceedings of the IEEE 72, no. 9 (1984), 1143–1163.

359. D. Gorenstein, **Finite Groups**. Chelsea, 1968.

360. L. C. Grove and C. T. Benson, **Finite Reflection Groups**. Springer-Verlag, 1984.

361. T. Hales, What is motivic measure? Bull. Amer. Math. Soc. 42, no. 2 (2005), 119–135.

362. B. Harfe, P. Scherz, S. Nissim, H. Tian, A. McMahon, and C. Tabin, Evidence for an expansion-based temporal Shh gradient in specifying vertebrate digit identities. Cell 118, no. 4 (2004), 517–528.

363. J. Hjelmslev, Neue Begrundung der ebenen Geometrie. Math. Ann. 64 (1907), 449–474.

364. G. Hjorth, Borel equivalence relations, in **Handbook on Set Theory** (M. Foreman and A. Kanamori, eds.). Springer, 2006. ISBN 978-1-4020-4843-2.

365. D. Hobby and R. McKenzie, **The Structure of Finite Algebras** (Contemporary Mathematics **76**). Amer. Math. Soc., Providence, RI, 1988.

366. J. H. Holland, **Adaptation in Natural and Artificial Systems**. University of Michigan Press, Ann Arbor, 1975.

367. J. E. Humphreys, **Reflection Groups and Coxeter Groups**. Cambridge University Press, 1990.

368. L. Kalmár, Zur Theorie der abstrakten Spiele. Acta Sci. Math. Szeged 4 (1928/29), 65–85. English translation in M. A. Dimand and R. W. Dimand, eds., **The Foundations of Game Theory**, Vol. I. Edward Elgar, Aldershot, 1997, pp. 247–262.

369. A. Kanamoru, Zermelo and set theory. Bull. Symbolic Logic 10, no. 4 (2004), 487–553.

370. I. Kapovich and P. Schupp, Genericity, the Arzhantseva-Olshanskii method and the isomorphism problem for one-relator groups. Math. Annalen 331, no. 1 (2005), 1–19; math.GR/0210307.

371. S. Katok, p-adic analysis in comparison with real, in **MASS Selecta** (S. Katok, A. Sossinsky, and S. Tabachnikov, eds.), Amer. Math. Soc., Providence, 2003, pp. 11-87.

372. D. Knuth, **The Art of Computer Programming. Vol. 2: Seminumerical Algorithms**. 3rd edition, Addison-Wesley, 1998.

373. N. Koblitz, **Algebraic Aspects of Cryptography**. Springer-Verlag, Berlin, 1998.

374. P. C. Kocher, Timing attacks on implementations of Diffie-Hellman, RSA, DSS, and other systems, in **Advances in Cryptography: Proceedings of Crypto '96**, Springer-Verlag, 1996, pp. 104–113.

375. P. C. Kocher, J. Jaffe, and B. Jun, Differential power analysis, in **Advances in Cryptography: Proc. of Crypto '99**, Springer-Verlag, Lect. Notes Comp. Sci. 1666 (1999), 388–397.

376. A. N. Kolmogorov, Local structure of turbulence in an incompressible fluid for very large Reynolds numbers. Doklady Acad Sci. USSR 31 (1941), 301–305.

377. D. König, Uber eine Schlussweise aus dem Endlichen ins Unendliche. Acta Sci. Math. Szeged 3 (1927), 121–130.

378. J. R. Koza, M. A. Keane, and M. J. Streeter, Evolving inventions. Scientific American, 288, no. 2 (2003), 44–59.

379. J. P. S. Kung, Combinatorics and nonparametric mathematics. Annals of Combinatorics 1 (1997), 105–106.

380. G. L. Litvinov, The Maslov dequantization, idempotent and tropical mathematics: A brief introduction, math.GM/0507014.

381. J.-L. Loday, Realization of the Stasheff polytope. Archiv der Mathematik 83 (2004), 267–278.

382. O. Loos, **Symmetric spaces. I: General theory**. W. A. Benjamin, Inc., New York-Amsterdam, 1969, viii+198 pp.

383. Yu. I. Manin, **Provable and Unprovable**. Sovetskoe Radio, Moscow, 1979 (in Russian).

384. Z. Michalewicz, **Genetic Algorithms + Data Structure = Evolution Programs** (3rd rev. and extended ed.) Springer-Verlag, Berlin, 1996.

385. G. Mikhalkin, Enumerative tropical geometry in $\mathbb{R}^2$. J. Amer. Math. Soc. 18, no. 2 (2005), 313–377.

386. J. Milnor, On the concept of an attractor. Communications in Mathematical Physics 99 (1985), 177–195.

387. M. Mitchell, **An introduction to Genetic Algorithms**. The MIT Press, Cambridge, MA, 1998.

388. P. Moszkowski, Longueur des involutions et classification des groupes de Coxeter finis. Séminaire Lotharingien de Combinatoire, B33j (1994).

389. R. J. Nowakowski, ed., **More Games of No Chance**. Cambridge University Press, Cambridge, 2002, xii + 536p. ISBN 0521808324.

390. R. Oaxaca, Male-female wage differentials in urban labor markets. International Economics Review 14 (1973), 693–709.

391. R. Oaxaca and M. Ransom, Calculation of approximate variances for wage decomposition differentials. J. Economic and Social Measurement 24, no. 1 (1998), 55–61.

392. A. Okabe, B. Boots, K. Sugihara, and S. N. Chiu. **Spatial Tessellations—Concepts and Applications of Voronoi Diagrams**. 2nd ed. John Wiley, 2000, 671 pp., ISBN 0-471-98635-6.

393. A. Yu. Ol'shanskii, Almost every group is hyperbolic. Internat. J. Algebra Comput. 2 (1992), 1–17.

394. P. P. Pálfy, Unary polinomials in algebras I. Algebra Universalis 18 (1984), 262–273.

395. P. P. Pálfy and P. Pudlak, Congruence lattices of finite algebras and intervals in subgroup lattices of finite groups. Algebra Universalis 10 (1980), 74–95.

396. J. Richter-Gebert, B. Sturmfels, and T. Theobald, First steps in tropical geometry, arXiv:math.AG/0306366.

397. S. Ricard, Open problems in Costas arrays, in **Seventh International Conference on Mathematics in Signal Processing, 18–20 Dec. 2006, The Royal Agricultural College, Cirencester**. The Institute of Mathematics and Its Applications, 2006, pp. 56–59.

398. R. T. Rockafellar, **Convex Analysis**. Princeton Univerity Press, 1970.

399. J. Schaeffer, N. Burch, Y. Björnsson, A. Kishimoto, M. Müller, R. Lake, P. Lu, and S. Sutphen, Checkers is solved. Science 317 (14 September 2007), 1518–1522. DOI 10.1126/science.1144079.

400. A. Schrijver, **Combinatorial Optimization. Polyhedra and Efficiency.** Vol. B, Springer, 2003.

401. E. M. Schröder, Eine grouppentheoretisch-geometrische Kennzeichnung der projective-metrischen Geometrien. J. Geometry 18 (1982), 57–69.

402. S. Shelah, Choiceless polynomial time logic: inability to express, in **Computer Science Logic 2000** (P. Clote and H. Schwichtenberg, eds.). Lect. Notes Comp. Sci. 1862 (2000), 72–125.

403. S.-J. Shin, Heterogeneous reasoning and its logic. The Bulletin of Symbolic Logic 10, no. 1 (2004), 86–106.

404. M. B. Skopenkov, Теорема от высотах треугольника и тождество Якоби. Математическое Просвещение 11 (2007), 89–99.

405. J. Snoeyink and J. Stolfi, Objects that cannot be taken apart with two hands. Discrete Comput. Geom. 12 (1994), 367–384. Available at `http://tinyurl.com/32o5yy`, accessed 9 July 2009.

406. R. P. Stanley, **Enumerative Combinatorics**. Vol. 2, Cambridge University Press, 1999.

407. R. P. Stanley, Solutions to Exercises on Catalan and Related Numbers, `http://www-math.mit.edu/~rstan/ec/catsol.pdf`, accessed 12 July 2009.

408. D. Speyer and B. Sturmfels, Tropical mathematics, `arXiv:math.CO/0408099`.

409. A. Szepietowski, **Turing Machines with Sublogarithmic Space**. Lect. Notes Comp. Sci. 843. Springer-Verlag, Berlin, 1994.

410. S. Thomas, Cayley graphs of finitely generated groups. Proc. Amer. Math. Soc. 134 (2006), 289–294.

411. K. Thompson, Retrograde analysis of certain endgames. ICCA Journal 9, no. 3 (1986), 131–139.

412. E. Tufte, **The Visual Display of Quantitative Information**. 1983: ISBN 0-9613921-0-X. Second edition: 2001.

413. A. Turing, The chemical basis of morphogenesis. Phil. Trans. Roy. Soc. London B 237 (1952), 37–72.

414. K. Vela Velupillai, Algorithmic foundations of *computable* general equilibrium theory. Applied Mathematics and Computation 179 (2006), 360–369.

415. E. B. Vinberg, Калейдоскопы и группы отражений. Математическое Просвещение 3, no. 7 (2003), 45–63.

416. G. Voronoi, Nouvelles applications des paramètres continus ła théorie des formes quadratiques. J. für die Reine und Angewandte Mathematik 133 (1907), 97–178.

417. E. Zermelo, Uber eine Anwendung der Mengenlehre auf die Theorie des Schachspiels, Proc. Fifth Congress Mathematicians, (Cambridge 1912), Cambridge University Press 1913, pp. 501–504. English translation can be found in U. Schwalbe and P. Walker, Zermelo and the early history of game theory, Games and Economic Behavior 34, no. 1 (2001), 123–137.

418. B. Zilber, Pseudo-exponentiation on algebraically closed fields of characteristic zero. Annals of Pure and Applied Logic 132, no. 1 (2005), 67–95.

419. B. M. Zlotnik (Б. М. Злотник), **Помехоустойчивые Коды в Системах связи**. Радио и Связь, Москва, 1989.

420. C. Zong, What is known about unit cubes. Bull. Amer. Math. Soc. 42, no. 2 (2005), 181–211.

421. C. Zong, **The Cube: A Window to Convex and Discrete Geometry**. Cambridge University Press, 2006. ISBN 0-521-85535-7.

Miscellaneous

422. C. Aldred, **Egyptian Art**. Thames and Hudson, 1980.

423. R. Arnheim, A comment on Rauschenbach's paper. Leonardo 16, no. 4 (Autumn 1983), 334–335.

424. J. B. Deregowski, Wire toys in Africa south of the Sahara. Leonardo, 13 no. 3 (1980), 207–208.

425. S. Eisenstein, The filmic fourth dimension, in **Film Form** (Jay Leyda, trans.). New York, Meridian Books, 1949.

426. **Etre et Avoir**, dir. Nicholas Philibert, France, 2002.

427. R. Feynman, **QED: The Strange Theory of Light and Matter**. Princeton University Press, 1985.

428. R. Gregory, **Mirrors in Mind**. W. H. Freeman, 1997, 302 pp.

429. L. Lehmann-Norquist, S. R. Jimerson, and A. Gaasch, **Teens Together Grief Support Group Curriculum. Adolescence Edition; Grades 7–12**, Psychology Press UK 2001, 165 pp., ISBN 1583913025.

430. *The Independent*, 5 May 2005.

431. G. Naylor, **Better Than Life**, New American Library, 1996. ISBN 0451452313.

432. D. Petty, **Origami, Paper Projects to Delight and Amaze**. D & S Books, ISBN 1-903327-08-3.

433. D. Petty, **Origami Wreaths and Rings**. Zenagraf, ISBN 0-9627254-1-2.

434. T. Pratchett, **Small Gods**. Transworld Publishers, London, 1993. ISBN 0-552-13890-8.

435. T. Pratchett and S. Briggs, **The Discworld Companion**. Vista, 1997. ISBN 0-575-60030-6.

436. B. V. Rauschenbach, On my concept of perceptual perspective that accounts for parallel and inverted perspective in pictorial art. Leonardo 16, no. 1 (1983), 28–30.

437. A. V. Voloshinov, "The Old Testament Trinity" of Andrey Rublyov: Geometry and philosophy, Leonardo 32, no. 2 (1999), 103–112.

438. O. Žholobov, Old Russian Counting: Девяносто, тридевять, четыремежидесяма, сорок. Russian Linguistics 28, no. 3 (2004), 409–416. DOI: 10.1007/s11185-004-1970-y.

Miscellaneous Internet Resources

439. M. Aston, Review of **Life without Genes: the History and Future of Genomes** by Adrian Woolfson, in the Balliol College Annual Record 2001. (Quoted from Peter Cameron's web page *Quotes on mathematics.*, `http://www.maths.qmul.ac.uk/~pjc/comb/quotes.html`, accessed 11 July 2009.)

440. Benezet Centre, `http://www.inference.phy.cam.ac.uk/sanjoy/benezet/`, accessed 12 July 2009.

441. Cal Sailing Club, **Introductory Handbook for Sailing Boats**. `http://www.cal-sailing.org/images/stories/files/dinghy09.pdf`, accessed 12 July 2009.

442. **On-Line Encyclopedia of Integer Sequences**, `http://www.research.att.com/~njas/sequences/Seis.html`, accessed 12 July 2009.

443. **Father Ted**, a cult TV sitcom (UK, Channel 4, Hat Trick Productions; written by Graham Linehan and Arthur Matthews). Official network homepage: `http://www.channel4.com/programmes/father-ted`, accessed 12 July 2009.

444. Introductory Assignment, Gelfand Correspondence Program in Mathematics, `http://gcpm.rutgers.edu/problems.html`, accessed 12 July 2009.

445. Houston Zoo Conservation Program, `http://www.houstonzoo.org/getFile.asp?File_Content\_ID=1100`.

446. MathSciNet, `http://www.ams.org/mathscinet`, is the Internet portal to Mathematical Reviews, a huge body of concise information about the world's mathematical research since 1940.

447. M. Minsky, Automated proof of Pons Asinorum, newsgroup post of 24 December 1999, `http://www.math.niu.edu/~rusin/known-math/99/minsky`, accessed 12 July 2009.

448. National Institutes for Health, `http://www.ncbi.nlm.nih.gov/`.

449. T. Relph, Counting sheep, `http://www.lakelanddialectsociety.org/counting_sheep.htm`, accessed 28 August 2008.

450. Royal National Institute for the Blind, Curriculum Close-Up 1: Maths, `http://www.rnib.org.uk/xpedio/groups/public/documents/publicwebsite/public_cu1t.txt`, accessed 12 July 2009.

451. Royal National Institute for the Blind, Curriculum Close-Up 13: Primary Maths, `http://www.rnib.org.uk/xpedio/groups/public/documents/publicwebsite/public_cu13w.doc`, accessed 12 July 2009.

452. W. Schwartz and K. Hanson, Equal mathematics education for female students, ERIC/CUE Digest, No. 78, Feb 1992.

453. **De Tribus Impostiribus**, `http://www.infidels.org/library/historical/unknown/three_impostors.html`, accessed 12 July 2009.

454. Trichotillomania Learning Center, `http://www.trich.org/`, accessed 12 July 2009.

455. V. A. Uspensky, Предварение для читателей "Нового литературного обозрения" к семиотическим посланиям Андрея Николаевича Колмогорова, `http://www.kolmogorov.pms.ru/uspensky-predvarenie.html`, accessed 12 July 2009.

456. V. A. Uspensky, Лермонтов, Колмогоров, женская логика и политкорректность, `http://www.kolmogorov.pms.ru/uspensky-lermontov_kolmogorov.html`, accessed 12 July 2009.
457. World Federation of National Mathematics Competitions, `http://www.wpr3.co.uk/wfnmc/info.html`, accessed 12 July 2009.

Index